A Philosophy of Technology

From Technical Artefacts to Sociotechnical Systems

A Philosophy of Technology - From Technical Artefacts to Sociotechnical Systems

Pieter Vermaas, Peter Kroes, Ibo van de Poel, Maarten Franssen, and Wybo Houkes

ISBN: 978-3-031-79969-3   paperback
ISBN: 978-3-031-79971-6   ebook

DOI 10.1007/978-3-031-79971-6

A Publication in the Springer series
SYNTHESIS LECTURES ON ENGINEERS, TECHNOLOGY, AND SOCIETY

Lecture #14
Series Editor: Caroline Baillie, *University of Western Australia*
Series ISSN
Synthesis Lectures on Engineers, Technology, and Society
Print 1933-3633   Electronic 1933-3641

Synthesis Lectures on Engineers, Technology, and Society

Editor

Caroline Baillie, *University of Western Australia*

The mission of this lecture series is to foster an understanding for engineers and scientists on the inclusive nature of their profession. The creation and proliferation of technologies needs to be inclusive as it has effects on all of humankind, regardless of national boundaries, socio-economic status, gender, race and ethnicity, or creed. The lectures will combine expertise in sociology, political economics, philosophy of science, history, engineering, engineering education, participatory research, development studies, sustainability, psychotherapy, policy studies, and epistemology. The lectures will be relevant to all engineers practicing in all parts of the world. Although written for practicing engineers and human resource trainers, it is expected that engineering, science and social science faculty in universities will find these publications an invaluable resource for students in the classroom and for further research. The goal of the series is to provide a platform for the publication of important and sometimes controversial lectures which will encourage discussion, reflection and further understanding.

The series editor will invite authors and encourage experts to recommend authors to write on a wide array of topics, focusing on the cause and effect relationships between engineers and technology, technologies and society and of society on technology and engineers. Topics will include, but are not limited to the following general areas; History of Engineering, Politics and the Engineer, Economics , Social Issues and Ethics, Women in Engineering, Creativity and Innovation, Knowledge Networks, Styles of Organization, Environmental Issues, Appropriate Technology

A Philosophy of Technology - From Technical Artefacts to Sociotechnical Systems
Pieter Vermaas, Peter Kroes, Ibo van de Poel, Maarten Franssen, and Wybo Houkes
2011

Tragedy in the Gulf: A Call for a New Engineering Ethic
George D. Catalano
2010

Humanitarian Engineering
Carl Mitcham and David Munoz
2010

Engineering and Sustainable Community Development
Juan Lucena, Jen Schneider, and Jon A. Leydens
2010

Needs and Feasibility: A Guide for Engineers in Community Projects — The Case of Waste for Life
Caroline Baillie, Eric Feinblatt, Thimothy Thamae, and Emily Berrington
2010

Engineering and Society: Working Towards Social Justice, Part I: Engineering and Society
Caroline Baillie and George Catalano
2009

Engineering and Society: Working Towards Social Justice, Part II: Decisions in the 21st Century
George Catalano and Caroline Baillie
2009

Engineering and Society: Working Towards Social Justice, Part III: Windows on Society
Caroline Baillie and George Catalano
2009

Engineering: Women and Leadership
Corri Zoli, Shobha Bhatia, Valerie Davidson, and Kelly Rusch
2008

Bridging the Gap Between Engineering and the Global World: A Case Study of the Coconut (Coir) Fiber Industry in Kerala, India
Shobha K. Bhatia and Jennifer L. Smith
2008

Engineering and Social Justice
Donna Riley
2008

Engineering, Poverty, and the Earth
George D. Catalano
2007

Engineers within a Local and Global Society
Caroline Baillie
2006

Globalization, Engineering, and Creativity
John Reader
2006

Engineering Ethics: Peace, Justice, and the Earth
George D. Catalano
2006

A Philosophy of Technology

From Technical Artefacts to Sociotechnical Systems

Pieter Vermaas, Peter Kroes, Ibo van de Poel, and Maarten Franssen
Delft University of Technology

Wybo Houkes
Eindhoven University of Technology

SYNTHESIS LECTURES ON ENGINEERS, TECHNOLOGY, AND SOCIETY #14

ABSTRACT

In *A Philosophy of Technology: from technical artefacts to sociotechnical systems*, technology is analysed from a series of different perspectives. The analysis starts by focussing on the most tangible products of technology, called technical artefacts, and then builds step-wise towards considering those artefacts within their context of use, and ultimately as embedded in encompassing sociotechnical systems that also include humans as operators and social rules like legislation. Philosophical characterisations are given of technical artefacts, their context of use and of sociotechnical systems. Analyses are presented of how technical artefacts are designed in engineering and what types of technological knowledge is involved in engineering. And the issue is considered how engineers and others can or cannot influence the development of technology. These characterisations are complemented by ethical analyses of the moral status of technical artefacts and the possibilities and impossibilities for engineers to influence this status when designing artefacts and the sociotechnical systems in which artefacts are embedded.

The running example in the book is aviation, where aeroplanes are examples of technical artefacts and the world aviation system is an example of a sociotechnical system. Issues related to the design of quiet aeroplane engines and the causes of aviation accidents are analysed for illustrating the moral status of designing, and the role of engineers therein.

KEYWORDS

philosophy of technology, technical artefacts, sociotechnical systems, ethics of technology, designing, use, operators, technological knowledge

Contents

Introduction

Technology is not just a collection of technological products, but it is also about how people, more in particular engineers, develop these products, about how people use them to meet their specific ends, and about how this all changes the world as we know it. Ultimately, technology is an expression of our endeavours to adapt the world in which we live to meet our needs and desires. Technological action may therefore be termed a form of goal-oriented human behaviour aimed at primarily resolving practical problems. Yet, the most tangible results of such efforts are the material products of technology, and we shall call them *technical artefacts*.[1] Technical artefacts are physical objects such as nails, aeroplanes and skyscrapers, but they are also the tools, machines and factories that are used to fabricate those objects. But practical problems are not just resolved by introducing a bunch of technical artefacts into the world. With these artefacts come instructions for their use. And with these technical artefacts come also social roles for people and social institutions for enabling the use of the artefacts. In this book, we shall present a philosophical analysis of this created technological world in its full splendour. We shall cover a wide spectrum of issues ranging from questions concerning the nature of technical artefacts to analyses of the methodology and knowledge underlying design, and ranging from characterisations of how technology extends into the social world, to the ethical evaluation of technological actions in personal and social fields.

We shall start by focussing on what exactly it is that distinguishes technical artefacts from other physical objects, such as those found, for instance, in the natural world that are not made by human beings. We shall also contemplate how technical artefacts are made, how they are designed by engineers, and what kind of knowledge this entails. Technical artefacts are used by humans to achieve all kinds of goals and that immediately gives rise to issues that involve more than just the artefacts. Through technology, people are able to change the world around them, and those changes, together with the accompanying goals, can be morally good or bad. In other words, technological questions are also ethical questions. Does technology help to make the world a better place or does it make the world more dangerous? To what extent can engineers be held responsible for all the good and bad consequences of technological developments? These are also issues that will be addressed. Through our discussions, we shall show how technology not only adds technical artefacts to our physical world but also enriches our social world with personal and social objectives, and with the actions required to achieve those objectives.

In this book, we shall make use of two main tools to guide the reader through this spectrum of issues. The first tool involves the concept of *use plan*. In short, use plans describe the ways in which we manipulate technical artefacts to achieve our different goals. One need only think of the instructions for use, accompanying any common-or-garden appliance, indicating precisely how it

---

[1] For a discussion of various forms of technology, see Mitcham, C. [1994].

can be used to, for instance, cut a hedge or listen to music. That, then, is the appliance's use plan. This first tool makes it possible for us to analyse how technical artefacts can become means to ends and allows us to see how there is a link between technology as a collection of tangible objects and technology as a collection of actions, by which people achieve their goals. In the first four chapters of these lectures, we shall be focusing predominantly on matters pertaining to the nature, the design and the ethics of individual technical artefacts. In so doing, we shall make plentiful use of this tool of use plans. In the chapters five to seven, the perspective will broaden out to include more social and societal technological issues, and for that purpose, we shall introduce a second tool in the form of the concept of *sociotechnical system*. Naturally, all technical artefacts, are embedded in human action, but in the case of many technical artefacts, they are also allied to wider systems where the proper functioning of the artefacts not only depends on the technology in question but also on social factors. Those kinds of systems are sociotechnical systems. An aeroplane factory and the civil aviation sector are examples of two sociotechnical systems, the first for producing planes and the second for the safe transporting of people by air. Such systems are composed of different kinds of elements including technical artefacts, people and also legislation which, together, form one cohesive system. For the individual user, technical artefacts embedded in sociotechnical systems are often difficult or impossible to use. However, with the help of all the individuals and legislation that constitute parts of such sociotechnical systems, users are often able to achieve their desired goals with these particular technical artefacts. With our two tools, we shall present what amounts to a modern view of technical artefacts, of the people who successfully adapt the world with those artefacts, and of the way in which those technical artefacts and people become socially and ethically intertwined.

The philosophy of technology touches on a wide range of issues, but there are also limits. Not all purposeful human action falls under the heading of technology, and we, too, will have to confine our areas of attention within the perimeters of this book. Science, for example, does not equate with technology. Admittedly scientists do impinge on the world with their experiments, all of which are conducted with specific objectives in mind, but scientists do not do that in order to adapt the world to fulfil various human needs and desires. Scientists intervene in the world in order to find better ways to observe and study matters and phenomena. As such, they are essentially world observers with an interest in knowledge. Similarly, in this book, political, economic or artistic actions are not considered to be part of technology. Government, for example, endeavours to directly influence the behaviour of people by drawing up and introducing different laws. Such laws have an impact on people's behaviour. Indeed, they change the legal landscape. In much the same way newly established companies alter the economic reality, and music, film and fashion all serve to enrich our cultural universe. In the very broadest sense of the word, one could say that all of these factors touch on technology, but that is not an area we shall be exploring here. Instead, the technology dealt with here will be primarily associated with how we change our material or physical world as we contemplate the designing and constructing or fabricating of, for example, bridges or microchips. In so far as political, social and legal matters are dealt with, it occurs mainly in relation to our discussions on ethical issues and on sociotechnical systems.

Finally, just a few words need to be said about the approach we take to philosophy in this book. Here philosophical questions are seen very much as questions that are posited in order to clarify concepts. In the case of technology, these are concepts such as: technical artefact, design, function, responsibility and, of course, the concept of technology itself. They are all central to the way we think about technology and its relationship to humans, but the meaning of these concepts is not always clear. In much the same way that engineers design material tools – technical artefacts – to get a better hold on the physical world, so we, philosophers of technology, need to develop conceptual tools in the form of concepts and conceptual frameworks, in order to get a better grasp on the way in which we think about technology.

It is a pleasure to acknowledge Malik Aleem Ahmed, Diane Butterman, Chunglin Kwa and Carl Mitcham for their various support. This book is an improved version of a text published in Dutch: *Vermaas, P., Kroes, P., Van de Poel, I., Franssen, M., and Houkes, W. 2009. Kernthema's in de Technische Wetenschap, Wetenschapsfilosofie in Context. Amsterdam: Boom.*

CHAPTER 1

Technical Artefacts

We start in this chapter by analysing the nature of technical artefacts. What sorts of objects are they? What are the typical characteristics of technical artefacts? Given that technical artefacts play such a predominant role in our modern world, these are all important questions. The way in which people interact with one another and relate to nature is determined, to a very high degree, by the technological resources they have at their disposal. In order to gain a better understanding of the nature of technical artefacts, we shall first compare them with natural and social objects. We shall also stop to consider the moral status of technical artefacts, which will involve us in the question whether they may be seen as good or bad, in the moral sense of the word. We will come to the conclusion that technical artefacts are physical objects designed by humans that have both a function and a use plan. If one adopts this view, then technical artefacts themselves can actually be morally significant.

1.1   A WORLD MADE BY HUMANS

Human intervention in the material world has taken on such huge proportions that one might safely assert that we live, nowadays, in a human-made world. Technical artefacts have come to dominate our environment and affect almost every facet of our behaviour. They have virtually superseded natural objects in every-day life, to the extent that we now even feel obliged to protect nature. The most familiar types of technical artefacts are common objects of use such as tables, chairs, pens, paper, telephones, calculators, et cetera. There is a huge variety of technical artefacts from very small to very big, from simple to complex, from component part to end-product and consisting of chemical materials, et cetera. What all of these things have in common is that they are material objects that have been deliberately produced by humans in order to fulfil some kind of practical function. They are often described as technical *artefacts* in order to emphasise that they are not naturally occurring objects. Here, we do not classify artistic works as *technical* artefacts because they do not fulfil any practical kind of function (the producing of art typically also draws on different creative skills and qualities than those demanded of an engineer).

The notion of technical artefacts and the distinction between natural and artificial objects presents many philosophical problems. Take the assertion that technical artefacts exist. It is indisputably true that humans enrich the world in which they live with all kinds of objects such as cars, calculators, telephones and so on. None of those objects existed a couple of centuries ago, but today they do, thanks to our technological ingenuity. What, though, do we precisely mean when we state that such technical artefacts exist? Technical artefacts are artificial (or man-made) objects. But the word 'artificial' also has the connotation of 'unreal' or 'not actual'. Does this mean that the very existence of technical artefacts, as artificial objects may be put into question? Take, for instance, a knife. For anyone who believes that the world is made up of only physical objects, in the sense that all the properties of objects can be fully described in terms of physics, a knife does not exist as *a knife*, that is, as a *technical artefact*, since 'knife' is not a physical concept. All that can be truly said to exist from that point of view is a collection of interacting particles – atoms and molecules – of which the knife is made up. Some metaphysicians therefore deny technical artefacts an existence of their own. For Peter van Inwagen, for instance, a knife is nothing more than atoms that are organised in a knife-like way.[2] Yet other metaphysicians, like Lynne Rudder Baker and Amie Thomasson do grant technical artefacts an existence of their own, separate from the atoms of which they are made up.[3] The distinction between artificial and natural objects is also problematic. It may be taken to be an artificial distinction in itself in the sense of being 'unreal', since one could point out that people themselves are natural organisms and that everything that a human being, as an integral part of nature, produces is inevitably natural (in much the same way that a beaver's dam may be termed a natural object). By attributing fundamental significance to the natural-artificial distinction, it is almost as though we humans are inclined to somehow view ourselves as external to nature. That, in turn, instantly gives rise to the question of how precisely humans may be viewed as non-natural beings.

These are not the kinds of questions that will preoccupy us here. In this chapter, we shall confine ourselves to a 'conceptual analysis' of the term 'technical artefact'. We shall endeavour to elucidate the term by comparing the way in which we describe technical artefacts with the way in which we describe natural and social objects (what exactly is meant by a social object will become clear in Section 1.3). It will be presumed that similarities and differences in the methods of description tell us something about the nature of technical artefacts. It is hoped that in that way, we will gain more insight into that nature, into how technical artefacts differ from other objects in the world, in particular from natural objects and social objects.

We take the modern aeroplane as an example of a technical artefact. Planes have now existed for more than a century, and they are deployed for various ends. Our focus will be on those used for civil aviation purposes, such as the Airbus A380. There are many different kinds of descriptions for such planes, ranging from very general descriptions of the type given in encyclopaedias and advertisements to extremely extensive technical descriptions of the kind given in, for instance, maintenance manuals. In those descriptions, one may distinguish three different aspects, each of which is important when answering the question of exactly what kind of an object a civil aviation aircraft is. The first aspect relates to the question of the *purpose* served by planes (for example, they are used for transporting people). The second aspect pertains more to the aeroplane's *structure* and to how it is built up (comprising a fuselage, wings, engines, a tail, a cockpit, and so on). Finally, the third aspect has to do with how a plane is *used* (describing, for example, what a pilot does to get it airborne).

[2] Van Inwagen, P. [1990].
[3] E.g., Baker, L. [2007] and Thomasson, A. [2007].

In general, there are always at least three relevant questions that can be asked about technical artefacts:

Question 1: What is it for?

Question 2: What does it consist of?

Question 3: How must it be used?

The answers to these questions will describe the following respective matters:

Aspect 1: The technical function of the technical artefact.

Aspect 2: The physical composition of the technical artefact.

Aspect 3: The instructions for use accompanying the technical artefact.

These three aspects are not independent of each other because the physical composition must be such that the technical function can be fulfilled, and if that function is to be realised, the user will have to carry out certain actions as described in the instructions for use as laid down in, say, the user manual. A technical artefact cannot therefore be considered in isolation of the accompanying instructions for use. These instructions specify a certain *use plan*,[4] a series of goal-directed actions to be met by the user with respect to the technical artefact to ensure that the function is realised, presuming that the technical artefact is not broken. Each technical artefact is, as it were, embedded in a use plan that is closely attuned to the relevant function.

In the light of what has been stated above, we may therefore define a technical artefact as *a physical object with a technical function and use plan designed and made by human beings.* The requirement that the object must be designed and made by humans is added to ensure that any natural objects that happen to be used for practical purposes are not also termed technical artefacts. For instance, a shell can be used for the purpose of drinking water. In such a case, it has a function and a use plan. Nevertheless, since it does not also meet the requirement of being human-made, it is not termed a technical artefact.

We shall now first consider how technical artefacts, as physical objects with a function and a use plan, differ from natural and social objects. We shall see that the first question (What is it for?) cannot be sensibly applied to natural objects and that the second question (What does it consist of?) has no relevance to social objects.

1.2   TECHNICAL ARTEFACTS AND NATURAL OBJECTS

The most obvious difference between natural objects and technical artefacts is the fact that the latter result from purposeful human action whilst the same cannot be said of natural objects; a natural forest is one that has spontaneously evolved, without any kind of human intervention. Nature is that which has in no way been subjected to or disrupted by any kind of human intervention. The essence of this difference between nature and technology, which is in many respects problematic, takes us back to a distinction first made by Aristotle between things that evolve through their very 'nature' and things which owe their origins to other causes.[5] Things that exist by nature – natural things –

---

[4]Houkes and Vermaas [2010].

[5]*Physica*, Book II, 192^b.

possess in themselves their principle of change. The nature of a thing is an intrinsic developmental principle directed at realising the aim or end of that same thing. The 'aim' of a birch seed is to develop into a fully-grown birch tree, and the principle of growth lies within the birch seed itself. A fully-grown birch tree, which is the result of (caused by) that growth principle, of the nature of the birch seed, is thus a natural thing.

A bed made from birch wood is not, by contrast, a natural object because the wood of the birch tree has no internal principle of change that is geared to turning it into a bed. A bed is not a naturally occurring object because it possesses, as bed, not an intrinsic principle of change. As Aristotle rightly observed, if you plant a piece of wood taken from a bed, it will not automatically grow (change into) a new bed. The reason for the existence of the bed lies outside the bed. It lies, of course, in its designer and maker who create it with a certain purpose in mind. The bed itself has no aim or end, and insofar as a bed has a function, it has that function only in relation to an external aim of the designer or user; here one sees, again, how close the relationship between function and use plan is. It is no natural object, but it is the result of the human skill (Τεχνή or *techné* in Greek); it is a technical artefact.

A bed, as a bed, therefore has no nature in the Aristotelian sense of the word. This is the reason that, over the course of time, a bed requires maintenance if it is to continue to fulfil its rightful function. Wear and tear and natural processes occurring in the material of which a bed is made, like rotting, can eventually lead to the damaging of the technical artefact. In general, terms, the very fact that certain objects require maintenance or repair is a strong indication that one is actually dealing with technical artefacts. From that point of view, biological objects such as gardens or dairy cattle are also technical artefacts. They are a direct result of human intervention and cannot remain in existence without continued human maintenance and care. Natural objects do not share the same need for maintenance and care. For those objects, it is rather the case that all forms of human intervention may be directly seen as having a disruptive effect on their very nature.

For Aristotle, not only living (biological) objects have a nature, but physical objects also have a nature; it is their intrinsic principle of motion that endeavours to realise the ends of physical objects. The motion principle of a heavy stone, for instance, is to gravitate to its natural place, and the natural place for all heavy bodies is in the centre of the universe.[6] Whenever one throws a stone up into the air, one forces it to make an unnatural upward motion when, in fact, the intrinsic principle of motion of any heavy stone is to perform a downward motion towards its natural place in the centre of the universe.

It is not easy to relate the Aristotelian concept of nature with the notion of nature that lies at the basis of our modern natural sciences. That is mainly because the idea that physical objects may conceivably have an intrinsic end is one that has been abandoned. Yet curiously enough, a core element of the Aristotelian notion is still present in the natural sciences. The natural sciences are devoted to the study of natural objects and phenomena, in other words, objects and phenomena whose properties are not human-made but are rather determined by physical, chemical and biological

---

[6]For Aristotle, the centre of the earth is also the centre of the universe.

laws. Those laws may be seen as the intrinsic principles of change for those objects and phenomena and thus as their nature (or as the natural laws governing them). Examples of such natural objects and phenomena are electrons, the chemical reactions that turn hydrogen and oxygen into water, hurricanes, and so forth. The properties and the behaviour of these objects and phenomena are not the result of direct human action.

With this modern view of natural objects in the back of our minds, in which physical objects are also included as natural objects, we return to our analysis of the similarities and differences between technical artefacts and natural objects. We will concentrate upon two categories of natural objects, namely physical and biological objects. First, we will compare technical artefacts with physical objects. The most striking difference between an aeroplane and an electron is that the first has a function and use plan whilst the second does not. Physical objects, such as an electron, have no function or use plan; there is no place for functions and use plans in the description of physical reality. This does not mean that electrons may not perform functions in technological equipment. It just means that from a physical point of view, such a function is irrelevant because that function has no consequences whatsoever for the properties and behaviour of an electron as a physical object. The function fulfilled by a plane, by contrast, is an essential property of that thing as a technical artefact: if we ignore the relevant function and the use plan, then we are merely left with a physical and thus natural object and not a technical artefact.

Perhaps, at first, it looks strange to say that an aeroplane is a natural object when one abstracts from its function and use plan because even if you simply view it as an object without a function, the plane is something that is fabricated by humans and thus an artefact. However, it is a natural object in the sense that all its properties (the mass, form, et cetera) and its behaviour may be traced back to the laws of physics and to the physical (natural) properties of the atoms and molecules of which it is composed. Whenever we take a random material object and contemplate and study it as a physical object, then the history of that same object becomes irrelevant. It is, for instance, irrelevant whether the object was manufactured for a certain purpose or that it has come about spontaneously. Neither does the technical function of that material object, indicating for what and how people may use it, alter any of its physical properties. Much the same can be said of physical phenomena. Many of the phenomena that physicists study nowadays do not occur 'in nature' but are instead produced by the physicists themselves in their laboratories. Nevertheless, they remain natural phenomena.

A second important difference between technical artefacts and physical objects, and one that is closely allied to the first, is that technical artefacts, unlike physical objects, lend themselves to normative claims.' This is a good/bad aeroplane' or 'This aeroplane should be in good working order' are sensible assertions, but the same cannot be said of assertions such as 'This is a good/bad electron' or 'This electron should work'. Normative assertions about technical artefacts indicate that they function to a greater or lesser degree or that they simply fail to fulfil their function. Normative claims about technical artefacts must be carefully differentiated from normative claims, pertaining to the way in which they are used. A technical artefact can be said to be incorrectly used in an instrumental sense, which amounts to the claim that the artefact is used in a way that does not correspond to its use plan (which typically means that the function is not realised). Alternatively, it can also be said to be wrongly used in a moral sense. In this second case, the normative claim is, however, not one about whether or not the use plan is followed, it rather is a claim about the goals that are realised with the technical artefact and the moral status of those goals.

Let us now consider biological objects. How can an aeroplane be said to differ from a bird 'in the wild'? Here, too, the answer seems obvious: a plane has a function, a bird does not. Yet the differences are more subtle and complex than in the case of physical objects. As a rule, it is true to say that biologists do not indeed attribute functions to plants and animals. What they do is to attribute functions to parts of plants, to organs of animal or to specific behavioural patterns. This is where we encounter natural objects and phenomena with a biological function. In a number of aspects, though, biological functions are clearly different from technical functions. In the first place, biological functions are usually ascribed to the component parts and behavioural patterns of biological organisms, not to the organism itself.[7] Technical functions are ascribed to the parts of technical artefacts but also to the technical artefacts in question as a whole. In the second place, the biological functions of organs, for instance, are not related to use plans as in the case of technical functions: a bird does not use its wings or have a use plan for its wings. A third point of departure (linking up to the first point) lies in the fact that it is impossible to make normative assertions about organisms as a whole. The wing of a bird may be said to be malfunctioning but not the bird itself, as a biological organism. To conclude, insofar as functions do arise in nature, these functions would appear to be different from technical functions.

Generally speaking, neither physical objects nor biological organisms have functions, which is why in the case of these kinds of objects, unlike in the case of technical artefacts, it does not make sense to pose the question 'What is it for?' The differences between technical artefacts and biological organisms diminish when, for practical purposes, specific varieties or organisms are cultivated or engineered by means of genetic manipulation (like for instance the Harvard OncoMouse of the 1980s that was developed for cancer research or, in more recent times, transgenic mice such as the 'super mouse' that has four times as much muscle mass as a normal mouse, thanks to a couple of alterations in its DNA). Just like technical artefacts, such organisms do have a technical function. They, therefore, more closely resemble technical artefacts and, consequently, just as with technical artefacts, endeavours are made to obtain patents for such organisms (natural organisms cannot be patented). In turn, such endeavours give rise to much controversy, which partly arises from the fact that, in cases of this sort, it is not so clear whether we are dealing with technical artefacts or simply with biological objects that have been modified by humans.

The foregoing brings us to a general remark about the distinction between technical artefacts and natural objects. There is no sharp 'natural' dividing line between both kinds of objects. Certain natural shells can, without in any way being changed by people, be used *as* drinking cups. However, the mere fact that a shell is used in such a way does not mean to say that it is a drinking cup or

---

[7]Ecologists attribute functions to plants and animals but only as components of ecosystems, and ecologists, again, do not attribute functions to the ecosystems as a whole.

a technical artefact. It is not a technical *artefact* because it is not a human-made material object. Undoubtedly, though, one could call it a technical object. A tree can be deliberately planted in a certain place so that when it reaches maturity, it serves to shade you from the sun. That does not make the tree a parasol, but it gives it a technical aspect. Similarly, it is difficult to say how much a stone has to be altered by humans for it to be turned into a hand axe, or a plant or animal for it to be termed a technical artefact. There is no unequivocal answer to that question. The dividing line between the natural and artificial worlds is a sliding scale; there is no clear-cut division between the two. Yet, that does not mean that there is no clear difference between paradigmatic examples of natural objects and technical artefacts. As we have seen, those differences do exist and, to sum up, those differences relate especially to the status of having a function and a use plan, and to the accompanying possibility of making normative assertions.

1.3   TECHNICAL ARTEFACTS AND SOCIAL OBJECTS

Also the social world we inhabit is, to a high degree, one of human making. Just as an aeroplane is the result of direct human intervention in the material world so are a new traffic rule or a new legal body like a firm the results of deliberate interventions in the social world. Like in the case of technical artefacts, such social 'objects' have a function. The primary function of a new traffic law is, for instance, that of regulating the rights and obligations of people participating in (public) traffic.

In this section, we shall compare technical artefacts with social objects. Such a comparison is not simple because of the wealth and variety of different kinds of social objects. A law, government, state, marriage, border, road user, driving licence, driving offence, traffic policeman, organisation, contract, and so on, are all part of social reality. Roughly speaking, all these things play a role in ruling the behaviour of humans, their mutual cooperation and the relationships between humans and social institutions. All of that is done by a fabric of formal and informal rules. Here the main focus will be on a certain kind of social objects, namely objects such as money, driving licences or passports. From a technological perspective, they are interesting because usually technology tends to play a prominent part in the producing of these objects. At a first glance, one might thus be tempted to categorise them as technical artefacts. But, as we shall see, they are not 'real' technical artefacts but rather social objects. We shall take as our example the ten-euro banknote. Its social function is that it constitutes legal tender – everyone can pay with it.

In order to explicate the difference between a ten-euro banknote and an aeroplane, we shall conduct the following thought experiment. Imagine that you are seated in a good functioning aeroplane, that is to say, if we use (or operate) the aeroplane well, then it will fly properly ('well' and 'properly' meaning: according to specifications). Then just imagine that you and your fellow passengers suddenly become convinced that the plane no longer works, that it is not fulfilling its function (without the good functioning specifications/criteria having in any way changed). What will be the consequences of that for the functioning of the aeroplane? Will it instantly no longer operate? Will it suddenly break down? Such conclusions seem absurd. Whether or not aeroplanes function (meet the specifications) is not something that depends on what the users or any other parties might happen to think but rather on the plane's physical properties since it is its physical structures that have a bearing on the aircraft's functioning.

In the case of the ten-euro note, matters are quite different. With that note, precisely the same fate could befall it as that which befell the Dutch ten-guilder bank note on 1st January 2002 (and the German ten-mark note, or the Italian 10.000 lira note, et cetera) when it lost its status as legal tender. Despite the fact that the physical properties of the note had remained unchanged, it simply lost its power to serve as legal tender from one day to the next (indeed, much the same may be said of a driving licence or passport that expires; without undergoing any physical change, it loses its function). Evidently, such means of payment do not depend, for the fulfilling of their function as legal tender, upon their physical properties. Naturally, such bank notes do, however, need to be provided with ultra modern security systems if forgery is to be avoided, and that is why technology is so vital when it comes to the producing of such monetary units. Nevertheless, the physical features which, from a practical point of view, are *necessary* if forgery is to be prevented are not in themselves *sufficient* to realise the function of legal tender.

Unlike an aeroplane then, a ten-euro bank note does not fulfil its function on the basis of its physical properties. On the basis of what does it perform its function? In a broad outline, the answer amounts to the following. If a ten-euro bank note is to fulfil its function as legal tender, it has to be generally accepted as legal tender. Exactly whether the latter actually holds is all a matter of whether or not people see it is being legal and official. In other words, it is all down to their believing that it is legal tender. If they do not view it as such, then the ten-euro note is unable to fulfil its legal tender function and becomes relegated to the status of being merely a valueless piece of 'paper'. It is therefore only on the grounds of *collective acceptance* that a ten-euro bank note is able to fulfil its function of legal tender. As soon as such collective acceptance disappears, it is no longer able to fulfil its function (as was seen to occur in the case of the Dutch ten guilder note when the euro was adopted).

We may therefore conclude that there is an important difference in the way in which technical and social objects fulfil their functions. Technical artefacts fulfil their function by virtue of their physical properties whilst social objects depend for their function upon their social/collective acceptation. In the case of social objects, such as money, the actual physical form taken is really immaterial which is why money may take very many different forms (from salt to digital information, e.g., 'zeros' and 'ones' printed on a chip card). In the case of a social object, such as money, the question 'Of what does it consist?' does not really make sense if one bears in mind that it does not depend for its function on a given physical manifestation. The foregoing explains why, when it comes to designing social objects (laws, institutes, rules, et cetera), it is vital to possess knowledge of people's social behaviour (including matters such as: what gives rise to social acceptation and how can that be promoted). By contrast, when it comes to designing new technical artefacts, knowledge of the physical phenomena is required.

Despite the fact that there is a crucial difference between the ways in which technical and social objects fulfil their functions, it is not always possible to unequivocally classify objects as being either technical or social. We shall further elucidate this point by taking the example of the problem of regulating vehicular traffic at a crossroads junction. That may be resolved in a purely 'social' fashion by endowing somebody with the function (status) of traffic policeman or woman and allowing them to direct the traffic (in the way frequently done in the early years of the car). Another social method is that of introducing traffic laws. Such methods only work properly if all or virtually every vehicle driver either recognises the authority of the traffic policeman or woman or abides by the traffic laws that have been laid down. In both cases, it is a form of collective acceptance that is required. The problem can also be resolved in a totally technological fashion (as in the way attempted through the developing of automatic vehicle guiding systems) by equipping cars and road junctions with information systems so that the traffic can be regulated without the intervention of any form of action taken by drivers or whomsoever. Collective acceptation is thus no longer required but rather technological systems that are completely reliable and possess the necessary physical properties.

Certain practical problems may therefore be tackled from either purely technological or purely social angles, but a mixture of the two approaches is also conceivable. The way that traffic lights operate is a case in point. When we describe the function of traffic lights by stating that they serve to regulate traffic at a crossroads, then it is clear that traffic lights can only fulfil that function if there is evidence of both the relevant technological and social systems functioning properly. The technological (or material) system comprises such components as switch boxes and lamps mounted on poles and positioned at junctions with red, amber and green colour filters sections. The switch boxes have to ensure that the lamps situated in the three different compartments go on and off in the correct sequence. Even if that technological system works perfectly well, one still cannot guarantee that the traffic light is really fulfilling its function. For the whole system to work perfectly well, road users also have to adhere to traffic regulations in which such matters as the significance of the red, amber and green lights are legally fixed. The function of the traffic light system may therefore only be said to have been fulfilled when both the technological and the social processes in question operate in the correct way (and are properly attuned to each other).

If one asks whether traffic lights, together with the rules concerning the behaviour of road users in response to the different colours, are essentially a technical artefact or a social object, there is no clear cut answer to that question. It is not entirely a technical artefact but by the same token it is not a straightforward social object either. It is a mixture of the two; traffics lights are an example of a sociotechnical system (see Chapter 5). Given that a solution to the regulation of traffic at crossroads may be found to a greater or lesser degree in the technological or social areas, it again seems that there is no sharp dividing line between technical and social objects. Here again a word of warning: even if there is a seamless transition between technical and social objects, the distinction has its significance. Even if a set of traffic lights cannot be unambiguously classified as either a technical or a social object, it is perfectly feasible to characterise different aspects of the whole as being either technological or social.

1.4   TECHNICAL FUNCTIONS

We have described technical artefacts as physical objects that have been designed and made by human beings and that have both a function and a use plan. Moreover, we have noted that the function bears some relationship to the physical structure of the technical artefact and to the use plan. Now we shall more closely examine the question of what a function is, and we shall do that by analysing just how engineers use the term 'function'.

In technological practice, there are two particular ways of describing technical artefacts that are important. Those ways are descriptions from a structural point of view and from a functional point of view. A structural description of a technical artefact simply describes it in terms of physical-chemical and geometrical properties. That is how a physical scientist, who knows nothing about the technical artefact's function, would describe the thing after having analysed it in detail (in answer to Question 2 in Section 1.1). A functional description looks, in contrast to a structural description, at what the technical artefact is intended for without saying anything about the physical-chemical properties (in answer to Question 1 above). A typical example of a structural description would, for instance, be: 'Object x has such and such mass, form, colour, and so on'; a typical functional description would be: 'Object x is for y' in which y is a relevant activity. Both kinds of descriptions are indispensable in technological practice, especially when it comes to the matter of designing technical artefacts. Schematically speaking, the designing of a technical artefact commences with a functional description of the object to be designed and ends with a structural description of that same object (see Chapter 2). A complete structural description has to be given if the technical artefact is to be subsequently produced.

Both descriptive methods are essential if a technical artefact is to be fully described. A structural description only describes a technical artefact from the physical object viewpoint; it does not take into consideration the functional properties whilst, conversely, a functional description omits all the structural features. They may not therefore be termed rival descriptive modes; they are complementary because they supplement each other.

Regarding the question as to what a technical function is, two interpretations can be broadly distinguished. The first interpretation closely links functions to the objectives of human actions, to what a technical artefact is expected to do, to 'the goal served'. If one goes into more detail, functions are then described in terms of '*black boxes*,' for which only the input and the output is described (see Figure 1.1). The function of the technical artefact, of the black box, is thus to transform input into output. One looks, as it were, at the technical artefact merely from the 'outside' and describes what precisely it should do. The function of the technical artefact can only be said to have been realised at the moment the goal is achieved. This is typically a user's view of technical artefacts and one in which functions are closely allied to use plans and the related reported goals that users have in mind. It is a view of functions that actually plays a crucial role in the early phases of the design process when all the functional requirements of the artefact that is to be designed are established. This is a thoroughly normative characterisation of a function; if the technical artefact, represented as a black box, fails to transform the input into output, then it may be said to function badly or not at all.

Input   Output

Black box

Figure 1.1: The black box interpretation of function.

The black box representation is insufficient for the designing, making, maintaining or repairing of technical artefacts, all of which belong to the engineer's central task. To that end, the black box must be opened and viewed from the inside. In designing, the black box still needs to be given content. What then comes to the fore is the link between functions and the physical properties and capacities of the technical artefact as a whole together with all its component parts. In these activities, we see that engineers often interpret a function as a desired physical property or capacity of the technical artefact. While functions are viewed from this internal perspective, the goals of users disappear from sight, and the emphasis comes to lie on the structural aspects of the technical artefact in question. As long as a technical artefact has a desired capacity, it fulfils its function regardless of whether the aims of human actions are realised (see also Section 2.1). Take note that although this second more internal functional description is linked to the structural description of the artefact, they are not the same. This second functional description is a description of *desired* structural properties of the artefact whereas a purely structural description focuses on structural properties that the artefact *actually* has.

Just like the structural and functional descriptive methods, both interpretations of functions are indispensable in technological practice. A technical function is inextricably intertwined with the physical qualities and capacities of a technical artefact; these qualities and capacities ensure that a function is realised in practice. And the function of a technical artefact and the artefact itself, cannot be detached from the use plans and from the goals of human actions. It is only in relation to those goals that technical artefacts have functions and that normative claims, pertaining, for example, to the good or bad functioning of technical artefacts, make any sense.

Since the functions of technical artefacts cannot be contemplated in isolation of the goals of human action and if one presumes that from the moral angle those goals can be evaluated as good or bad, one may well wonder whether technical artefacts can, in themselves, be seen as morally good or bad. We are not talking here about technical artefacts being good or bad in an instrumental sense, that is, about technical artefacts realising their technical function; we are bringing up the question of whether it is meaningful to assert that they are good or bad in a moral respect. This is where we come up against the problem of the moral status of technical artefacts.

1.5   THE MORAL STATUS OF TECHNICAL ARTEFACTS

'*Guns don't kill people, people kill people.*' This slogan, once produced by the American *National Rifle Association,* is perhaps the most succinct way of summarising what is known as the neutrality thesis of technical artefacts.[8] What this thesis asserts is that from a moral point of view a technical artefact is a neutral instrument that can only be put to good or bad use, that is to say, used for morally good or bad ends, when it falls into the hands of human beings. People can use weapons to defend themselves from robberies but also to mount armed robberies. The weapon in itself can never be qualified as either good or bad in the moral sense of the word. The way in which the neutrality thesis specifically comes to the fore is in the notion of the 'dual use' of technical artefacts. It is a term that is used to indicate that technologies that can be used for peaceful ends, can often equally be deployed for military purposes. A case in point is radar systems that can be used to follow aeroplanes. On the one hand, radar can be used to make civil aviation safer, but they can equally be used in times of war to track and shoot down enemy aircraft.

The neutrality thesis has direct consequences for our views about the moral responsibility of engineers as designers and as makers of technical artefacts. The thesis asserts that technical artefacts do not in themselves have any moral implications, thus implying that engineers are involved in merely designing or making morally neutral instruments or means. As engineers can generally bring no influence to bear upon the way in which the technical artefacts they design are actually put to use, they cannot therefore be held morally responsible for the way in which their artefacts are ultimately used.

There are various arguments that can be levelled against the neutrality thesis, a few of which we shall briefly examine. In the first place, it may be remarked that in the design phase, engineers already anticipate not only what form their product might take but also how it may be used. In other words, they do not just design something that can be randomly put to good or bad use, but they design rather for a specific use. The whole idea that during the design phase engineers anticipate use is something that, for instance, emerges from the use plan notion. Whenever engineers design any artefact, they simultaneously contemplate the goals of the kind of user they have in mind, who will subsequently have to be able to fit that artefact into a certain use plan. Such use plans are invariably not morally neutral. A macabre example of a morally reprehensible use plan is that of the gas chambers of the Second World War in which a vast number of Jews were killed. In such circumstances, it is hard to defend the argument that such purpose-designed gas chambers can be construed as a neutral instrument that only became morally horrible in the hands of the Nazis.

In reply to that kind of argument, one might claim that whilst engineers are perhaps able to anticipate certain kinds of uses, they cannot in fact *determine* them. An artefact can always be put to different use to that which was originally intended, be misused or simply not used at all. Though this may be true, this does not mean to say that a designed artefact is always suitable for any given form of use. A radar system that has been designed for civil aviation purposes has to meet other requirements and will possess other technical features to one designed for military ends. Not all

---

[8]For a defence of the neutrality thesis, see Pitt, J. [2000].

techniques can be easily implemented for dual use. Certain techniques are often so designed that it is presumed, stipulated or tempting to believe that they will be used in a certain way. A good example of this is the traffic slowing down devices in the form of speed bumps. Such speed bumps are there to 'encourage' traffic participants to slow down in an endeavour to raise traffic safety levels. In this case, a certain degree of moral behaviour may also be said to have been built into the design of the technology in question. Obviously, there is good reason why, in English, speed bumps are also sometimes alternatively termed '*sleeping policemen*'. If such kinds of behavioural influencing can be built into the technology, we then speak in terms of *scripts*. In Chapter 3, we shall examine the moral aspects of scripts in a little more detail.

A second argument against the neutrality thesis lies in the fact that the phases of designing and using cannot always be clearly differentiated. This applies especially to situations in which sociotechnical systems, such as the civil aviation system, come into play. The functioning of sociotechnical systems is not only dependent upon the proper operating of technological systems but also upon properly functioning social systems (see Chapter 5). A company such as Airbus designs a component of this system, namely the aeroplanes, whilst the existing civil aviation system is fully in use, and this system is one of the elements that determine the constraints for the planes that are to be designed. Sometimes sociotechnical systems can be so all-embracing that the 'use' thereof becomes almost unavoidable, like, for instance, in the case of the electricity system. The built-up environment or a city, which may also be seen as a sociotechnical system, has become so commonly accepted in everyday life that it would be strange to speak in such cases of 'use'. Likewise, in the case of sociotechnical systems, the notion of design can sometimes also be problematic; all kinds of components of sociotechnical systems are designed, but often, there is no single organisation that is responsible for designing the system as a whole. On top of everything else, the way in which use is effected can sometimes also change the system itself. In these kinds of cases, where the dividing line between use and design becomes hazier, it is difficult to sustain the thesis that technology is only invested with morality when it is put to use.

A third argument that can be levelled against the neutrality thesis resides in the fact that new technical artefacts invariably lead to new options for acting, in other words, to ways of acting that did not previously exist and which could give rise to ethical questions. For instance, the invention of the aeroplane made it possible to speedily transport goods to distant destinations and that, in turn, gave rise to new moral obligations when it came to the matter of dealing with disasters and famine in distant regions. From time to time, new options for acting that are opened up by technology can be very controversial, precisely because of the inherent moral considerations – like in the case of the cloning of animals and people – and can therefore constitute grounds for wishing to oppose certain technologies.

In such cases, supporters of the neutrality thesis could argue that it is up to the users to either avail themselves of the new options for acting or not. However, in certain situations, the mere existence of new options to act is, in itself, something that is morally relevant. A good example of this is prenatal testing of the sort that can indicate whether an embryo has some kind of defect.

The detection of such defects can be reason enough for certain people to decide for an abortion, a decision that may be questioned morally. However, apart from any moral issues related to abortion, the mere possibility of carrying out a prenatal test will force parents to make a morally relevant decision, namely to either go for testing or not. In these kinds of cases, it is simply the creating of new options for acting that is morally loaded.

In our final argument against the neutrality thesis, we might point out that technical artefacts do not just fulfil functions but that they also bring with them a whole host of undesired side-effects and risks. The use of aeroplanes leads, for example, to noise hindrance, environmental pollution and sometimes even to accidents in which people are killed. Such kinds of side effects and risks clearly have moral significance. Adherents to the neutrality thesis could assert that also these side effects primarily depend upon the manner of use. That is not, however, always the case. The amount of noise hindrance created by an aeroplane does not, for instance, only depend on how it is used but also upon how it is designed. The occurrence of side effects also indicates that the designing of technical artefacts is not only about their efficacy and efficiency. One should also bear in mind that it is not just through being used for a certain purpose that technical artefacts influence the world but also through their side-effects. Such side-effects have to be accounted for in the design phase. The sorts of issues one may think of in this connection are safety, health, sustainability and privacy. These are all moral values that can already, as it were, be built into technical artefacts in the design phase. If such moral values are already inherent to technical artefacts, then one could level this as yet another argument against the neutrality thesis.

All in all, proponents of the neutrality thesis are very much inclined to detach the functions of technical artefacts from the specific aims and objectives of human dealings and to conceive of technical artefacts as objects with particular physical properties or capacities. In other words, they conceive of technical artefacts as physical objects. Though it is true to say that these physical objects are designed and made by people with a view to their particular physical properties, it is also true to assert that just like the physical properties of a natural pebble or electron these particular physical properties cannot be evaluated as good or bad in a moral sense. In other words, technical artefacts in themselves cannot be seen as either good or bad. Given such a view of technical artefacts, the neutrality thesis may be said to be applicable to them. However, if we think of technical artefacts as physical objects that have been designed and made by human beings and that have both a function and a use plan, as we propose, then the neutrality thesis can no longer be said to hold. The function and the use plan link technical artefacts inextricably to human goals, and since such goals have moral significance, the same has to be said of the technical artefacts to which they are related.

1.6   CONCLUSION: THE DUAL NATURE OF TECHNICAL ARTEFACTS

Our endeavours to conceptually analyse the notion 'technical artefact' have resulted in the following three key notions: 'physical object', 'function' and 'use plan'. The characterisation of an object as a technical artefact has to refer to a physical object, a function and a use plan (symbolised in Figure 1.2

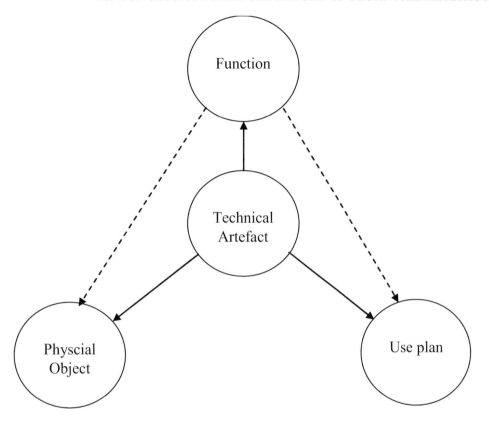

Figure 1.2: A conceptual anatomy of the notion of technical artefact.

by the arrows with continuous lines). We have furthermore established that the function of a technical artefact is, on the one hand, related to the physical object and, on the other hand, to the use plan (symbolised by the arrows with dotted lines). We have also concluded that technical artefacts are not morally neutral because their functions and use plans pertain to the objectives of human actions, and those actions are always morally relevant.

Our conceptual anatomy of the notion of technical artefact leads us to the conclusion that technical artefacts must be a very special kind of objects. We have already established that technical artefacts are different from physical (natural) objects and social objects. According to the interpretation given above, technical artefacts are hybrid objects that incorporate characteristics of both physical and social objects. An aeroplane is, on the one hand, a physical object with all kinds of physical features and capacities required for fulfilling its function. On the other hand, though, the function of a plane may not be termed a purely physical feature because it also pertains to a use plan or, in more general terms, to a context of human action. In that context of human action, goals have a crucial role to play, and it is only in relation to those goals that physical objects can be said to have functions. Just like the functions of social objects, the functions of technical artefacts are related to the purposeful (intentional) actions of people, but they cannot be termed social objects because the realisation of technical functions is something that comes about in a completely different way. To conclude, it may be asserted that technical artefacts have a *dual nature*:[9] they are objects that belong both in the world of physical (i.e., natural) objects and in the world of social objects.

1.7   A FEW MORE ISSUES

We have compared and contrasted technical artefacts with natural objects and social objects. Other comparisons are also possible, so we would ask you to consider the following questions. What is the difference between technical artefacts and waste, such as the sawdust that is created when wood is sawn or the carbon dioxide produced from burning? What is the difference between technical artefacts and art forms such as sculpting? What is the difference between technical artefacts and chemical substances? Is there, for instance, a difference between artificial vitamins that are chemically produced and the natural vitamins that are obtained from fruit and plants? With the aid of the terms 'physical object', 'function' and 'use plan,' you can clarify your answers to these particular questions.

A question that is more difficult to answer is the question of whether the objects that animals produce can be described as technical artefacts. You may perhaps be inclined to maintain that certain objects are technical artefacts like, for instance, the dams created by beavers or the twigs fashioned by apes to get ants out of anthills. Simultaneously, other products, such as cobwebs, might seem less likely candidates. Where will you draw the line? Is it possible to draw a vague line? Does it make sense to assert that animals allow objects to *slightly* fulfil functions by implementing use plans for those objects?

---

[9]Kroes and Meijers [2006].

CHAPTER 2

Technical Designing

In Chapter 1, a technical artefact was defined as a physical object designed and made by humans with a technical function and a use plan. In this chapter, we focus on the question of what technical designing actually is. There is consensus amongst engineers about two main features:

- the core activity of technical designing is describing a physical object that is able to fulfil a technical function effectively and efficiently;

- in a broad sense, technical designing is an activity aimed at achieving the goals of people, companies or society as a whole.

A philosophical reflection that connects with the practice of technical designing will begin with an analysis of these two characteristics. Such an analysis is, however, immediately thwarted by an ambiguity that we encountered in Chapter 1. Engineers interpret the concept of technical function in two different ways, which means that an analysis of the first mentioned characteristic of designing immediately derails in uncertainty. This complication simultaneously provides us freedom. We, as philosophers, are able to choose one of these interpretations, and we can do so in a way that helps us giving more interesting answers to questions about technical designing and about technical artefacts.

2.1   CHARACTERISING TECHNICAL DESIGNING

In the engineering literature, technical designing is characterised as an activity in which engineers describe a physical object that is capable of effectively and efficiently fulfilling a technical function. That technical function originates from a requesting party, the 'customer', which is typically a company but can also be a private individual or a government organisation. This customer may determine the required function himself or herself, or the function is deduced by the designing engineer from the goals, problems or wishes the customer presents. The physical object that is described is frequently an object that does not as yet exist but still has to be designed and made. Such an object is then, by definition, a technical artefact. A customer may, for instance, wish to transport people over long distances in short periods of time. From this stated objective, the engineer concludes that it is all about 'transporting x number of people over y distance in z period of time' before then going on to describe the physical structure of a technical artefact – a new aeroplane – that can fulfil this function. The artefact must be effective, meaning that the engineer should be able to justify that it actually fulfils the function in question. Furthermore, the artefact must be efficient in the sense that no reasonable better or cheaper alternatives can be found that meet the needs of the customer as well.

Let us formulate this characterisation of technical designing as follows: technical designing is an activity in which for a technical function F a description D_S of a technical artefact with a physical structure S is established that can effectively and efficiently fulfil F. There are at least two remarks that can be made about this characterisation. In the first place, engineers warn us that technical designing entails more than simply finding a description D_S of an artefact that can fulfil the function F. This characterisation of technical designing is therefore better be to seen as one that captures the *core activity* of technical designing. In the second place engineers do not use the term technical function in an unequivocal way (see Section 1.4). The characterisation of the core activity of technical designing itself cannot thus be said to be completely unequivocal. We shall return to this second point later in this section.

If describing a physical structure S of an artefact that can fulfil a function F is merely the core activity of technical designing, then the question comes up what forms more extensive characterisations will take. A first elaboration follows from the observation that the decision of engineers to opt for a certain physical structure S for the artefact that is to fulfil a function F will all depend on the current state of the art in technology. In its turn, the current technological state of the art imposes restrictions on the precise choice of structure S and the exact functions F that can be fulfilled. The designing engineer will therefore sometimes have to consult the customer in order to ensure that the objectives of the customer correspond with what is technically possible. A designing engineer is not only a service provider who satisfies customers' requests but also an advisor who interprets, adapts and even sometimes rejects customer demands. A customer who wants to transport people at speeds of over 20,000 kilometres per hour will be told by engineers that that is impossible. If you wish to transport people at speeds of 2000 kilometres per hour, you will be told that – though possible – that amounts to an expensive and technologically complex option, just think of Concorde. The desire to transport people at speeds of 800 kilometres per hour will automatically translate into the function of 'carrying people by air' as it is technologically far less efficient to realise such a speed over land or by sea. It is part of designing, in this more elaborate characterisation, that the engineer exercises influence over the function F and such influence can also occur during the core activity of designing if, for instance, it becomes apparent that an aircraft type with an average cruising speed of 750 kilometres per hour proves to be considerably cheaper than one that cruises at 800 kilometres per hour.

Technical designing thus extends to a phase that precedes the core activity of establishing a description D_S of an artefact that can fulfil a function F. It also extends to the phase that follows. The description D_S is a description of an artefact for which the designer is able to technically justify that it can fulfil F. This justification may sometimes exist of giving a demonstration that a prototype artefact with structure S indeed is able to perform the desired function F. Yet, in cases where the artefact then has to be produced in great numbers – think of car manufacture – the design process can further extend to, for instance, the necessary manufacturing facilities, including the technological infrastructure required for maintenance and repair.

Methodologically, and in terms of timeframes, the core activity of designing is not, moreover, isolated but rather intertwined through all kinds of feedback loops with the preceding and ensuing phases. If the description D_S of an artefact with function F is impossible or too complex, the designing engineer can return to the customer and ask if he or she can adapt his or her goals. But also after the description D_S of an artefact has been laid down or after it has been proved that the prototype works, it is still possible for the engineer to go back to the customer and modify the goals. The artefact can be difficult to manufacture, or it can turn out that the fabrication costs are too high. In some cases, it is only once an aeroplane has been manufactured or has even entered into service that a clear view of the production costs emerges; one need only think in this context of the new problems that continuously surface during Space Shuttle flights. Designing is therefore an *iterative* activity or, in other words, an activity in which, step by step, engineers translate the goals of customers into functions F, convert those functions into descriptions D_S of artefacts, construct those same artefacts, test them and ultimately produce them, while at every new stage being able to return to previous steps.

In the engineering literature, technical designing is sometimes placed in an even broader context, which brings us to the second feature of designing mentioned at the beginning of this chapter. The customers are frequently commercial companies – many engineers are traditionally employed by such companies – which means that the vested interests of those companies, in the form of commercial success and economic continuation, become integral aspects of the goals that have to be achieved during the design process. Technical designing thus also becomes, in part, a form of product design for economic markets; the designing engineer must create artefacts that can be marketed and must innovate technologically so that companies and national economies can grow. As well as this commercial aspect of technical designing, there is also a social aspect that needs to be considered. Governments also commission projects: for instance, in order to serve society engineers design airports, air traffic control systems and military aircraft. Engineers are service providers who set out to help people, companies and society to realise their goals and to resolve their problems. Engineers thus have great economic and social influence both in positive and negative ways. People can travel by air, but aeroplanes can also crash and have a major impact on, for example, those who live near airports and upon the environment. As such, technical designing is also becoming, to an increasing degree, a process of designing for demands and needs in the areas of ethics, safety, the environment and sustainability.

This wealth of different characterisations makes philosophical reflections on technical designing similarly multifaceted. In the present chapter, we shall provide a philosophical reconstruction of technical designing where the main focus throughout will be on the core activity. In the next chapter, we shall go on to discuss the ethical aspects of technical designing.

One problem that still needs to be addressed is the fact that the characterisation of the core activity of technical designing cannot be precise as long as the meaning of the term 'technical function' has not been fully defined. In Section 1.4, we distinguished two interpretations: one in which functions are closely related to the goals of human actions and one in which functions are interpreted as the desired physical properties or capacities of technical artefacts. The two interpretations are different, which can be established if one questions under what circumstances the artefact has fulfilled its function. In the case of a plane, for instance, the first interpretation leads to take the goal of transporting people by air from point a to point b as the plane's function. In the second interpretation, the function of the plane may be described as that of having the capacity to transport people through the air. With the first interpretation, the aeroplane may be said to have fulfilled its function as soon as the goal is achieved, that is, once the aircraft has landed and the passengers have arrived at their destination; in the second case, the plane's function is fulfilled as soon as it becomes airborne with passengers on board (regardless of whether it ever reaches its destination). It is precisely this ambiguity in the concept of function that extends to ambiguity in the characterisation of technical designing. With the first interpretation, the core activity describes an artefact that is able to attain a possible goal whereas in the second interpretation, it is an artefact that has certain desired physical properties or capacities. In some cases, the difference between these two specifications of designing may become irrelevant, like when it becomes the goal of the designers to design an artefact with a given kind of physical capacity, such as a quieter aeroplane engine. In other cases, though, the difference can be important. It may be argued that, for engineers, there are many more possibilities if the primary intention behind the artefact they design is to transport people over long distances in a short span of time rather than to design an artefact that merely has the capacity to transport people by air; in the first case, a high speed train could also emerge as a possible outcome of the design process.

When analysing how engineers define functions and how they characterise technical designing, this kind of ambiguity is a fact that simply has to be noted and accepted. But when reflecting philosophically on technology, it is better to decide exactly how we are going to implement the term function. We, as philosophers, can adopt an unequivocal meaning by deciding how we are going to define the term 'technical function'. We shall opt for the second interpretation, and we shall systematically take technical functions to be the desired capacities of artefacts. By making this choice, we obviously disregard that part of the engineering literature in which functions are taken to denote goals. Engineers who conceive of functions as goals may therefore view the remaining analysis of designing given here as strongly biased. There are two reasons for supporting the choice that we propose. To start off with, a philosophical analysis requires that concepts have a clear and unambiguous meaning. We therefore have to make a choice, even if that means estranging ourselves, in part, from the discipline that we set out to describe. In the second place, our choice enables us to examine technical designing in more detail. For engineers who maintain that functions are closely related to goals, the core activity of designing consists of just one overall step from a goal to a description of an artefact with which that goal can be achieved. But when we take functions as desired capacities, we can divide up this one step into sub steps: in technical designing, the goal is first converted into a description of the desired physical features or capacities of technical artefacts, and then into a description D_S of an artefact that has those properties or capacities. (For engineers who think of functions as desired capacities, only the second step is their core activity; for such engineers the first step belongs to technical designing in the broader sense of the word).

With our decision to regard functions as the desired physical properties and capacities of artefacts, we thus identify a step in technical designing in which objectives are converted into functions (as desired physical properties and capacities). We shall now describe the relationship between these goals and functions with the aid of a use plan, as outlined in Section 1.1. A use plan consists of the actions that the user of an artefact has to fulfil with the artefact in order to realise the goal associated with that particular artefact. Such a use plan is fairly clear for light aircraft and consists of actions such as 'go and sit in the plane', 'start the engine', 'taxi at high speed along a straight and level runway' and 'adjust the position of the flaps on the wings as soon as the craft is airborne'. In the case of civil aviation aircraft, the use plan is much more complicated and, apart from involving flying aspects, it also involves actions pertaining to flight controllers, passengers and their baggage. For instance, with flight controllers, decisions have to be made regarding the flight route, the passengers have to be seated and switch off their mobile phones, and the baggage has to be checked for explosives. In Chapter 5, which is devoted to sociotechnical systems, we shall return to the matter of the designing of technological systems in which people are also involved. The relationship between the use plan and the function of the artefact lies in the fact that the function is the physical property or capacity that the artefact must have, according to the engineer, for the use plan to be effective. The use plan for a small plane amounts to a series of consecutive actions aimed at transporting people by air from a to b, and the function of the aircraft lies in the capacity to carry people in the air, since this is the capacity of the aircraft which ensures that the execution of the use plan realises the goal of the plan.

In the use plan analysis, the core activity of technical designing consists of four steps:

1. the designing engineer commences with fixing the customer's goal,

2. (s)he then develops a set of actions with the artefact-to-be-designed by which this goal can be realised,

3. (s)he determines which function F, that is, which capacity F, the artefact must have if the plan with the artefact is to indeed realise the customer's goal, and

4. (s)he finally describes the physical structure D_S of the artefact so that it has that capacity F.

This kind of use plan analysis will generally not be found in the relevant engineering literature.[10] Engineers may point to the importance of compiling a good manual explaining how to correctly use the designed artefact and manuals may be seen as descriptions of use plans. Yet, in reality, engineers mainly seem to view manuals as necessary by-products of their work, whilst in the use plan analysis manuals occupy as descriptions of the use plans a central position in the process of technical designing. The use plan analysis therefore has the status of a *rational reconstruction* of technical designing. It is

---

[10]One exception being Roozenburg and Eekels [1995, Section 4.3].

an analysis that adds extra structure and detail to the engineer's characterisations. Because of this addition of extra structure, the analysis has a somewhat arbitrary character; alternative reconstructions are conceivable, which leads to question what precisely the benefits are of adding the concept of a use plan to an analysis of technical designing. We shall explain this benefit. At the end of this chapter, we shall use the use plan analysis to describe how technical designing is connected to uses and to designing by people who are not engineers. And in the next chapter, we will use use plans in our ethical reflections on artefacts and technical designing.

2.2   THE STRUCTURE OF TECHNICAL DESIGNING

Because of the broader contexts in which the core activity of technical designing is embedded, it is not so that every description D_S of an artefact that is able to fulfil a function F is automatically a case of successful technical designing. Those contexts impose all other kinds of further restrictions. The function F must derive from the goals of a customer and that same customer can have further demands, such as physical and financial requirements and demands linked to the use of the artefact that is to be designed. The social context also brings with it limitations. When being produced, used and disposed of, artefacts must comply with various safety and environmental norms and stipulations. These restrictions must all be taken into consideration when describing D_S of the artefact that is to be designed. The designing engineer must furthermore complete the assignment within a given timeframe and with given resources: not only the engineer but also the facilities that he or she makes use of, cost money. The technical designing of an aeroplane must therefore be completed within a stipulated span of time, and the aircraft must comply with certain physical requirements – the width must, for instance, be adapted to meet standard aeroplane dimensions – financial demands must be considered, the costs of construction and maintenance must fit within the customer's budget and safety, ergonomic and environmental norms must all be accounted for. All in all, these factors mean that technical designing is a very different ball game to performing, for example, scientific research or to doing more explorative research into new technological possibilities. When it comes to technical designing, it is necessary to be sufficiently certain that the description D_S of an artefact that is able to fulfil the relevant function F can actually be produced within a limited span of time, with the resources available and on the basis of the available technological means. The designing of an aeroplane must be feasible. By contrast, scientific and technological research may remain uncertain in terms of expense and success. When formulating a new chemical theory, doing archaeological excavations or carrying out research into nuclear fusion, one is not, in principle, expected to prove beforehand that success is guaranteed and that results can speedily be produced.

This is the reason why, to a large extent, technical designing remains a process of *re*designing: the technical designing of a plane, for instance, often amounts to introducing a variation in existing designs so that, in that way, success may be guaranteed within the permitted design period. Walter Vincenti, an aeronautical researcher, calls such redesigning *normal* designing, and he contrasts it with *radical* designing. In normal designing, engineers adopt the *operational principle* and the *normal* *configuration* of an artefact from existing designs.[11] According to Vincenti, the operational principle is the physical and technological principle that enables the artefact to fulfil the required function. The normal configuration, by contrast, is the way in which the salient components of the artefact are ordered. For present generations of commercial aircraft, the operational principle resides in the creation of upwards lift via wings, and the normal configuration consists of a fuselage in which the passengers sit with two central, sideways protruding wings, two small side tail wings and one upright tail stabiliser. Radical designing is any kind of designing where engineers deviate from the operational principle, from the normal configuration or from both. In many cases, technical designing amounts to normal designing; radical designing is often confined to just several components of an artefact such as when introducing new materials or making various engine innovations.

One reason why engineers often want to stick to redesigning is because a great deal of time and energy has to be put into the analysing and checking side of matters. In technical designing there are three main phases that can be distinguished: a *conceptual phase* in which the configuration of the artefact and the most important components are described in general functional terms, a *materialisation phase* in which technical solutions in terms of existing technical components are selected for the various components and a *detailing phase* in which the physical description of the chosen components is adapted so that all the demands can be met. Especially the last two phases are rounded off with necessary but time-consuming analyses to determine whether or not the results achieved are adequate. The time and means demanded by these analyses are all at the expense of the time and means that the engineer is able to devote to designing in the conceptual phase. This is especially the case for radical designing since, then, less use can be made of existing designs for analysis purposes. In the various engineering literature, much attention is nowadays devoted to the conceptual phase because that is when technological innovations – radical designs – can be achieved. This shift in attention can partly be attributed to the increased use of computers, for thanks to computers, it has become much easier to consult databases to find out about the properties of existing designs, and so the whole process of analysing has become far less time-consuming.

The conceptual phase is sometimes described as a phase in which the designing engineer 'decomposes' the intended function F in an ordered series of subfunctions $f_1, f_2, \ldots f_n$ without at the same time immediately opting for existing components for fulfilling those subfunctions. In that way, the engineer is able to dissect and analyse the intended function without immediately committing himself to existing operational principles and normal configurations. The function 'transporting people by air,' can thus be broken down into the subfunctions 'giving people a seat' and 'conveying by air' without immediately having to envisage seats in a circular tube-like structure with wings, engines and landing gear. Alternative solutions – a helicopter, a zeppelin, a flying saucer? – are thus not, by definition, ruled out. The selection of components is something that then takes place in the materialisation phase, when it becomes clear whether the existing components suffice and whether some new components have to be designed. In the detailing phase, the descriptions of the selected and newly defined components are refined so that all the original requirements can be met.

---

[11] Vincenti, W. [1990].

To summarise, the core of technical designing lies in finding a description D_S of an artefact with a physical structure S that is able to effectively and efficiently fulfil a certain function F. If this description is given on the basis of existing operational principles and normal configurations for artefacts with F, then one may speak of normal technical designing or of redesigning. If other operational principles and/or normal configurations are chosen than existing ones, one is dealing with cases of radical technical designing or of innovative designing. Both kinds of technical designing can be divided up into three phases: a conceptual phase in which the configuration of the artefact and the important components are functionally described, for instance, by breaking down the function F into subfunctions $f_1, f_2, \dots f_n$; a materialisation phase in which technical components are chosen for the various parts, and a detailing phase in which the physical descriptions of the components are refined.

2.3   REASONING IN TECHNICAL DESIGNING

We can now take up the issue of the way in which engineers reason when technically designing, and analyse it in some detail on the basis of the analysis provided in the last section. The matter of the knowledge implemented by engineers during that process is the topic of discussion in Chapter 4.

The conversion of a technical function F into a physical description D_S of an artefact is not perceived by engineers as a logical derivation. Engineers presume that for every function F, there exist different physical structures $S, S', \dots$ for artefacts that can fulfil F. So, then there are also numerous possible descriptions $D_S, D_{S'}, \dots$ Conversely, it is also presumed that an artefact with a given physical structure S can fulfil a number of functions $F, F', \dots$ Both presumptions can be covered by the slogans *Function doesn't follow form* and *Form doesn't follow function*. The consequence of this is that reasoning in technical designing is not a deduction of the description D_S from the given function F. This reasoning has more the character of a *decision*: the designing engineer analyses possible descriptions $D_S, D_{S'}, \dots$ of artefacts that can fulfil F and then takes a technologically supported decision to subsequently select one as the most effective and efficient description.

On the basis of the different distinctions introduced in the last section, it is possible to further analyse the decision-making character of technical designing. We can furthermore begin to indicate upon what grounds the various decisions are taken (though we cannot yet give a complete analysis of these grounds). The first decision that a designing engineer has to take is whether it is normal or radical designing that is most suitable for describing an artefact that can fulfil a certain particular function. The advantage of opting for normal designing is that it makes the design process efficient: such a decision largely anchors the decisions that have to be taken in the conceptual phase as the engineer is immediately able to base his or her further plans on an existing operational principle and an existing normal configuration. The configuration and its most important parts are thereby quickly determined. In the materialisation phase, the engineer has to decide once again. In this phase, there is a transition from the functional description of parts in terms of subfunctions to physical descriptions of technological components that can fulfil those subfunctions. For that transition, the designing engineer can make use of existing knowledge about existing components and the functions that they can fulfil. Opting for normal designing does not actually imply that an engineer will be mimicking every single detail of certain previous designs. Existing designs are often protected by patents but can invariably be improved upon in the light of new-found techniques. One constraint in the area of component selection emanates from what engineers call *integration*: the selected components must technologically match. When designing an aeroplane, it would not, for example, be very sensible to opt for engines that run on *kerosene* and for electricity generated by a *diesel* aggregate for the plane's various instruments. The materialisation phase, leads to an initial description D_S of the structure of the artefact to be designed. In the detailing phase, an analysis is then made to establish whether this description suffices by determining if an artefact described by D_S is able to fulfil the function F while also meeting the other requirements that have been laid down. Within normal designing, use can be made of existing technical knowledge of the chosen components: the weight-bearing capacity of existing wing types is, for instance, known so that by varying the different parameters of the wing – the length, breadth, form, et cetera – a specific wing can be created so that the plane that has to be designed meets the requirements that have been laid down.

Within normal designing, the main task of the engineer is to focus on the materialisation and detailing phases; within radical designing, the conceptual phase is also of importance. There can also be underlying economic reasons for opting for radical designing – a company might want an aeroplane that is innovative so that it can in that way create or capture a new market. Alternatively, it might want a plane that is technically advanced because the operating principles and normal configurations of the current aircraft perhaps no longer meet the possibly stricter safety and environmental requirements. In the conceptual phase, the engineer has then to lay down a new configuration or possibly think up a new operational principle and reason starting from the established function F to an ordered series of subfunctions $f_1, f_2, \ldots f_n$. This line of reasoning is also essentially more of a decision than a derivation since engineers presume that there are a number of possible ordered series of subfunctions that are capable of realising a given function F. Within radical designing, the materialisation and detailing phases again follow the conceptual phase, at which point, it is then possible that the knowledge of existing components is proved to be useless: the chosen ordered series of subfunctions $f_1, f_2, \ldots f_n$ may include subfunctions that cannot be fulfilled by existing components so that new components have to be developed for the subfunctions in question. The designing of such new components could hail the dawn of a new technical design project in which the requested subfunction is further broken down into new subfunctions for which existing components could be used.

In design methodology, various methods have been developed that can guide engineers in their reasoning processes throughout the different phases. We shall discuss just a few of them here. For the breaking down of functions F into ordered series of subfunctions $f_1, f_2, \ldots f_n$, Gerhard Pahl and Wolfgang Beitz and, more recently, Robert Stone and Kristin Wood, have devised a method, that is known as *functional modelling*.[12] In this particular method, functions and subfunctions are presented as operations on *flows of material, energy* and *signals*. The function F that is to be analysed

---

[12] Pahl and Beitz [1996, Section 2.1] and Stone and Wood [2000].

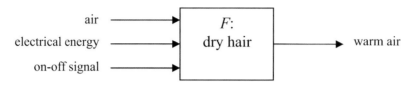

Figure 2.1: The function F of a hairdryer.

is represented as an operation that the customer wishes to see carried out upon such flows. Here the aeroplane example is not so suitable. We shall therefore consider a hairdryer. The function of such an artefact is to be represented as an operation by which incoming air, electrical energy and a signal from an on-off signal is transformed into a flow of warm air (see Figure 2.1). The subfunctions f_1, $f_2, \ldots f_n$ are represented by means of well-defined elementary technical operations, working on flows of material, energy and signals such as 'transport electrical energy', 'convert electrical energy to rotational energy', and 'stop acoustic energy'. In the conceptual phase, the engineer then has to select a network of such elementary subfunctions $f_1, f_2, \ldots f_n$ that, together, exert the same net operation on the flows as the overall function F. For the hairdryer, these subfunctions are 'transport electrical energy in accordance with an on-off signal', 'separate electrical energy into two flows of electrical energy', 'convert electrical energy to kinetic energy' and 'convert electrical energy to thermal energy' (see Figure 2.2).

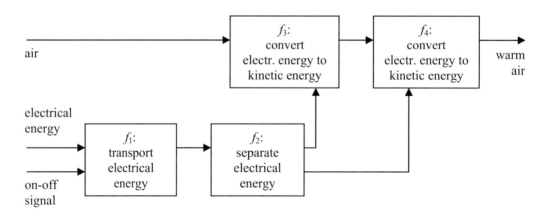

Figure 2.2: Subfunctions f_1, f_2, f_3 and f_4 of a hairdryer.

The idea underlying this functional modelling method is that in the materialisation phase the designing engineer can select existing components for each subfunction or sometimes even for a whole cluster of subfunctions. In more general terms, it is presumed in many design methods that engineers have a whole repertoire of existing components at their disposal – through experience, manuals and, more recently, computer data files – of which the subfunction are known. This repertoire is then plundered in the materialisation phase when determining which components can best meet the subfunctions f_1, f_2, ... f_n. In that way, designers are able to create different combinations of components that are able to fulfil the function F and thus also generate and assess a number of possible descriptions D_S for the artefact that is to be designed. One conclusion that can be drawn from this using and reusing of knowledge and experience, is that the description D_S is typically not derived from the function F on the basis of theoretical knowledge. Engineers do not, for instance, calculate the description D_S by filling in the function F in a theoretical or mathematical formula. It is rather the case that engineers make use of the knowledge and experience they have of the functions and physical properties of existing artefacts and components and so determine the descriptions D_S.

The materialisation phase thus leads to a description D_S of the artefact to be designed that consists of physical descriptions of the components and the way in which they are linked. This description is, not usually the ultimate description. In the detailing phase, the engineer further adapts the description so that the designed artefact meets all the previously laid down requirements. This final adapting may sometimes be compared to the resolving of a mathematical problem, namely of an optimisation problem in which the different variables of the components and the connections between them are so adapted that D_S meets the requirements. In the case of an aeroplane, the variables can, for example, consist of the number of passenger seats n, the weight of the aircraft w and the power of the engines p, together with a maximum consumption of x litres of kerosene per passenger per kilometre. In the detailing phase, the designer can then slightly adjust the aeroplane variables n, w and p in order to meet the requirements and to, in an ideal situation, manufacture an aeroplane that consumes the optimal minimum amount of fuel per passenger kilometre. A more systematic description of this type of reasoning in the field of technical designing would lead us to the methods of Herbert Simon.[13] According to his methods, engineers must translate design problems into mathematical optimalisation problems. The goals – the functions and other demands – of the customer must be translated into quantitative conditions with which the artefact must comply. The description of the artefact has to be parameterised in terms of design variables. A mathematical function has to be defined for a measure of the success to which the design, characterised in terms of the values for the different design variables, meets the conditions. In the example, the mathematical success function for every trio of values n, w and p for passenger seats, aircraft weight and engine power, will indicate just how much kerosene the aeroplane consumes per passenger and per kilometre. After that, the mathematical space of all the possible values for the design variables will be searched so that a choice of the values that are suitable for the variables can be made. Suitable means that, with the selected values, the artefact can at least be said to meet the requirements. It also means that there are no other values close to the selected ones that would make the artefact meet the requirements better. However, finding a suitable match for the values for the design variables does not mean that the values chosen constitute an optimal choice, that is to say, the choice that ensures that the design

[13] Simon, H. [1967].

meets the requirements as well as possible. Due to the size of the range of possible values for the design variables and due to the fact that engineers have to come up with a design under the pressure of limited resources and time, it is generally impossible to consider all the possible values. Therefore, an aeroplane characterised by certain values n, w, and p, for which the fuel consumption level per passenger and per kilometre lies below the x litre requirement level, will be deemed suitable, provided that small variations in the chosen n, w and p, values do not prove to leading to a more economical plane. Not all the n, w and p values will, however, be checked; precisely when it is a good time to stop searching for better alternatives is not something that can be unequivocally determined but which must be decided by the engineer. Simon called this search for a suitable, rather than an optimal, match for the design values 'satisficing'.

In technical designing, reasoning on the basis of optimalisation is a suitable method, provided that the goals and the descriptions of the artefacts to be designed have been sufficiently well laid down to be analysable in quantitative mathematical terms. The problem in technical designing is that many design problems do not fit neatly into that particular category. Design problems rather have to be converted by engineers into this mathematical form, for instance, by passing through the conceptualisation and materialisation phases. Again, such a conversion cannot be taken as a derivation but depends rather on decisions made in the conceptual and materialisation phases. In turn, these decisions can influence the design problem or the way in which the engineer interprets that problem. If, in those phases, an engineer opts, for instance, for a conventional aeroplane, then the stated safety demands have to be interpreted differently from the case in which the engineer opts for a zeppelin. Such a decision clearly has an influence on the quantitative modelling of the artefact that is to be designed. This influence of the decisions of engineers and the way they formulate the design problem is central to the *reflective practice* design method of Donald Schön.[14] According to that method, technical designing starts with an unstructured problem. The designing engineer then imposes order on this problem by 'naming' and 'framing' the problem in a certain way. In naming the problem, the engineer selects the elements he thinks are most important in the design problem, and through framing, he interprets that problem in a particular way. The vague problem of wanting to transport people quickly is, for instance, named as a problem about transporting people by air, and it is framed as the problem of needing to create an aircraft that can take off vertically, that can fly horizontally and can descend vertically. After that, the engineer takes the next step in the technical designing process by exploring solutions to the more structured problem he created, and then evaluates the results together with the adequacy of the chosen way of labelling and interpreting the problem. The engineer might perhaps choose a gas balloon as a means of providing upward force for the aircraft and then go on to establish whether that decision helps to simplify the remaining design problems. This evaluation will lead to new insights about the original design problem and the chosen solution. The design problem provides 'feedback' – Schön's term is 'back talk' – to which the engineer 'listens' for identifying better options to name and frame the design problem. Ultimately, the gas balloon proves to be too big and too unwieldy a solution for lifting people into the air in

[14]Schön, D. [1983, 1992].

an urbanised environment, which means that the problem has to be reinterpreted as the need to create a flexible kind of aircraft. In that way, a cycle is created in which the design problem and, in any case, the interpretation of that problem changes and different kinds of new solutions are investigated until the engineer is finally satisfied with the interpretation of the problem and with the chosen solution. In Schön's method, it seems, therefore, that the barrier between the conceptual and the materialisation phase is abandoned in favour of a freer, more creative but less definable way of reasoning.

2.4   ENGINEERS, DESIGNERS, USERS AND OTHER ROLES

In Section 2.2, we distinguished between technical designing and scientific research. In this section, we shall show how technical designing relates to the use of artefacts and also to activities that can be seen as technical designing but not as (technical) designing by engineers. In so doing, we shall again make use of the use plan analysis.

If technical designing merely consists of the core activity of translating functions into descriptions of artefacts, then it would appear that engineers predominantly serve their customers by furnishing them with artefacts: the customers go to the engineers with their goals and subsequently obtain from the engineers descriptions of artefacts that will help them to realise their goals. This way of typifying the relation between engineers and customers or users falls short in a number of ways. For instance, usually customers are not so much interested in descriptions of technical artefacts; instead, they are interested in the technical artefacts themselves. As an engineer, one is also expected to inform users about the actions that have to be performed with the artefacts if the goals in question are to be realised. And users frequently, and equally importantly, need to be informed of just what they must *not* do with the artefacts. Users have to be trained just how to fly a newly designed aeroplane, and they must be taught precisely which actions will endanger the passengers or people situated in the plane's vicinity at any given point in time.

This information conveyance from engineer to user has a very obvious normative side to it that is inevitably linked to any discussion surrounding the *proper* and *improper use* of artefacts. Commercial aeroplanes are designed to transport people from one place to another and may not be used to attack people and structures in the way demonstrated on 11th September 2001. This latter example makes it clear that the *improper* use of artefacts can be deeply reprehensible, but that is not always the case. Many of us sometimes use screwdrivers improperly, for instance, to ease open the lids on paint tins or to poke holes in materials. This certainly is not in line with the purposes for which screwdrivers were originally designed, but in many cases the alternative use is handy and rational.

With the help of the use plan analysis, it is possible to describe the relationship between designing engineers and users more clearly. In this analysis, the technical design process amounts to an activity in which the designer – typically an engineer – develops a new use plan and describes the artefacts that are manipulated in the process of carrying out that plan. The engineer serves his or her customers by supplying that use plan and the artefacts that are to be manipulated. The provision of information to the customer explaining how he or she can realise his or her goals with the aid of the artefacts is thus immediately described by means of the use plan. In the use plan analysis, the using of an artefact amounts, in its turn, to the executing of a use plan for an artefact so that the aim of the use plan can be realised. Someone has a goal and via information obtained from others – like parents (for everyday information such as how to prepare a meal), friends, advertising and schooling (for more advanced goals like internetting and flying) – is already familiar with a whole host of use plans for the purpose of realising goals. One puts those plans into practice, all depending on the available means such as time, experience and, of course, the artefacts themselves. What amounts to the *proper use* of an artefact is the execution of a use plan for that artefact that is *socially recognised*. Use plans are socially recognised if they have, for example, been developed by engineers because those are the people recognised as the relevant experts, as people who have been educated as experts in the field of use planning. Use plans for aeroplanes that have been developed by engineers are thus socially recognised. But there are also other ways in which use plans can gain social recognition. More everyday artefacts, such as kitchen utensils, have use plans that have existed for a long time and which have been passed on from generation to generation. Undoubtedly, such plans came into being without the intervention of engineers or other experts. Nevertheless, because of the widespread knowledge of these use plans, they have become just as socially recognised as the use plans developed by engineers. By contrast, *improperly using* an artefact has to do with the executing of a use plan that is either not recognised or not yet socially recognised. Transporting people by plane by manipulating the aeroplane in accordance with the instructions for use amounts to proper use but using the plane to blow up buildings amounts to improper use because it deviates from the use plan developed by engineers. Furthermore, if you cut your food with a knife, then that amounts to proper use, but if you tighten screws with a knife, that constitutes improper use.

With the use plan analysis, it is possible to distinguish more roles that people may have in relation to artefacts than just those of technically designing them and using them properly or improperly. In the first place, according to this analysis, an engineer also designs when he or she develops a specific use plan without having to describe also new artefacts. The engineer can serve a customer by producing a new use plan, which is designed for the manipulating of artefacts and natural objects that already exist. One need only think, for instance, of advice possibly provided by an engineer to use existing software packages to achieve certain goals. Such advice can be developed using the same methods as those applied in technical designing and can, thus, be technologically just as well substantiated; the only difference is that now the engineer merely selects the artefacts that are subsequently to be manipulated. This alternative role of selector can more or less be characterised with the term *consultant*; in the field of consultancy, problems can be resolved on the basis of existing means. In the second place, in the use plan analysis, there is also room for the development of use plans by people who are not engineers. Engineers do not, after all, have a monopoly on the development of use plans or on describing or making artefacts. In the analysis of proper and improper use, we referred above to use plans that have gained traditional acceptance such as the use plans that hold for kitchen utensils. Such use plans have a long history and originate from long before the period when we started distinguishing engineers as professional technical designers. If such a use plan is thought up by a specific person, then it would seem fairly obvious to also call that same person a designer. In more general terms, any individual who comes up with a use plan for a newly thought up artefact, for an existing artefact or for a natural object, may be perceived as a designer. If you mock up a new kind of kite for a child or help a fellow camper by fastening a loose guy rope with a screwdriver, then you become involved in technical designing. Such kinds of designing could be labelled *amateur designing* or, where they involve existing artefacts and objects, *innovative using*. Such kinds of designing differ from technical designing by engineers in that amateur designers or innovative users do not develop their new use plans and the accompanying artefacts with professional knowledge drawn from modern technology but rather, perhaps, on the basis of everyday experience. In the Middle Ages and before that, these forms of designing constituted the only possible way of arriving at new use plans; in our modern technological age, non-engineers gain less recognition. If, in our day and age, a non-engineer devises a new use plan for an existing artefact, it will often initially be seen as improper use of that artefact (one need only think of medicines such as Viagra that are used as party-drugs). As soon as anyone who is not a qualified engineer endeavours to describe a new artefact, it will often be difficult for that same person to prove that the artefact really does 'work'.

The advantage of the use plan analysis is that it provides a richer view of the roles that people can have in relation to artefacts. We can describe the relationship that exists between technical designing and using, and we can distinguish technical designing from other forms of designing such as amateur designing and innovative using. In so doing, we must recognise that notably innovative using, in general, will not be seen as a form of designing: for engineers and probably also for the users themselves, innovative using is merely viewed as a type of using. In this respect, the use plan analysis once again becomes a reconstruction of technical designing rather than a painstaking and realistic analysis; since technical designing is reconstructed as the development of a use plan for an artefact, innovative using that consists of the thinking up of a new use plan for an existing object also becomes a kind of technical designing. Note that in case the existing object, is a natural object, the use plan is not a use plan for a technical artefact since we defined technical artefacts as objects designed and made by humans.

With this richer view that is generated by use plan analysis, it also becomes possible to throw light on the discrepancy between proper and improper use of artefacts. In the case of modern technical artefacts, such as aeroplanes, this distinction can be traced to the *division of labour* that exists between engineers and non-technical, unschooled designers. Proper use of modern artefacts amounts to uses according to use plans that are developed on the basis of professional technological knowledge – with all the guarantees of success in the case of uses that engineers can give – and improper use consists of uses of the same artefacts according to use plans developed by non-engineers on the basis of other types of knowledge – with all the possible dangers that that might entail. With modern technical artefacts it would appear that thanks to the above-mentioned division of labour, engineers have a monopoly on use plans and on the describing of technical artefacts, so that innovative use of such artefacts or unintended use is quickly branded by engineers as 'improper use' with 'unintended consequences'. Civil aeroplanes are incorrectly used as projectiles. Low-energy light bulbs, designed to encourage people to use less energy, are often actually used to create more light sources so that people unintentionally end up using more energy rather than less. Nuclear energy is not merely a source of electricity, but it is something that could also, according to some, unintentionally lead to the creation of police states because the use of nuclear material has to be strictly monitored. A more positive description of the role of improper use of artefacts can also be given. Non-engineers have always had the capacity to develop new ways of using artefacts and that also applies to modern technical artefacts. Engineers do not have a monopoly on technical designing and thus on the determining of what constitutes the proper use of such artefacts, but they can improve their work by anticipating how users will manipulate modern artefacts by taking that aspect into account during the technical design phase.

2.5   CONCLUSION

In this chapter, technical designing has been reconstructed with the help of the notion of use plan. Technical designing is characterised as a process in which engineers develop use plans so that customers and/or users can realise given goals. Use plans involve actions with artefacts and, if they do not yet exist, those artefacts are in designing also described by engineers. At the same time, the describing of such artefacts defines the core activity of technical designing, which is to establish a description D_S of an artefact with a physical structure S that is effectively and efficiently able to fulfil a function F. In this reconstruction, the function F is taken to be the physical capacity that the artefact must have if the execution of the use plan is to realise the given goal.

The core activity of technical designing can be divided up into three phases: a conceptual phase in which the configuration of the artefact and the most important components are functionally described, a materialisation phase in which technological components are selected for the different parts and a detailing phase in which the physical description of the components is refined. In the case of normal designing, use is made of existing operational principles and normal configurations for artefacts with function F; in the case of radical designing, existing operational principles and/or normal configurations are deviated from.

With the use plan analysis, further roles are defined in relation to artefacts. Using has to do with the executing of use plans while consultancy is the developing, by engineers, of new use plans for existing artefacts and objects. Amateur designing and innovative use, on the other hand, has to do with the developing by non-engineers of new use plans for artefacts and objects. Proper use is analysed as the execution of socially recognised use plans, and improper use is conversely seen as the exercising of non-socially recognised use plans.

In the next chapter, we shall continue by looking at the ethical questions surrounding technical designing, and in Chapter 4, we shall return to the matter of technological knowledge. In these chapters the reconstruction of technical designing will be used and supplemented. Only in Chapter 5, when we extend the analysis of technical artefacts to include sociotechnical systems, will the use plan analysis be elaborated.

2.6   A FEW MORE ISSUES

The use plan analysis would appear to lend itself well to the developing of more technologically complex matters such as electronic products, machines and industrial installations. In the case of such artefacts, manuals or perhaps even courses, are normal necessary aspects of the design process and the ways and means by which users can learn the use plans. In the case of simple technical artefacts, however, such as wire, bouncing balls and door handles, it seems somewhat strange to think in terms of use plans that designers have to draw up for users. How can the design and use of such simple products be described on the basis of the use plan analysis?

A more difficult and controversial question is the one surrounding *Intelligent Design*. Adherents of creationism assert that many biological organisms have a level of complexity that cannot be entirely explained purely on the basis of evolutionary theory. According to them, this level of complexity can, however, be described as being designed. A question you might well like to consider is whether, in this case, designing means the same as technical designing. Are there customers, are there goals and are there use plans?

CHAPTER 3

Ethics and Designing

In the previous chapters, we discussed what technical artefacts are and how we can best understand the technical design process. We learned that technology is directed at changing the world and that engineers contribute to that by designing artefacts. The changes brought about by technology can be both for the good and for the bad. In this chapter, we shall therefore consider some of the ethical questions related to technical artefacts and designing. Ethics has to do with good and bad and right and wrong behaviour. Ethics is about how the world should be and not about how it actually is. Given that engineers are concerned with changing the world when they design technical artefacts, one might say that they are doing ethics. An important question about ethics in engineering is then: how can and should engineers change the world for the better by the designing of artefacts? The emphasis in this chapter lies on how engineers can deal with ethical questions during the design process. The relevant analysis takes as point of departure, the idea worked out in the previous two chapters that engineers not only design artefacts but also use plans. In this chapter, we shall presume that engineers are reasonably able to anticipate the impact their designs will have on society; in Chapter 7, we shall consider situations where that is not the case, in other words, where we are confronted with the unintended consequences of technology.

3.1   ETHICS

Three central notions in the area of ethics are values, norms and virtues. *Values* are enduring convictions about what is good; values can pertain to a good life, to a good or just society but also to what constitutes a good work of art. Values have to be differentiated from interests or preferences. Preferences or interests indicate what someone considers important for himself or herself; when it comes to values though an appeal always is made – at least implicitly – to something that applies to everyone. If we state that freedom is a value, then we do not merely mean to say 'I like to be free,' but we mean rather that is important for everyone to be and feel free. You can like floral wallpaper without having to believe that others should share the same view; in such a case, we speak about a preference rather than a value. Values are, moreover, *enduring* convictions. They do not change from one day to the next. That does not mean to say that values cannot change or that we cannot arrive at new insights about values; however, in such cases, more is at stake than just a change in our preferences. Rather than stating that we now like something else, we will, for instance, put forward arguments for why a value has become or is no longer of importance, or has to be differently understood.

Not all values are of a moral nature. Instances of non-moral values that are also important in the design process are beauty and simplicity. Alongside the distinction between moral and non-moral values, a distinction is often made between intrinsic and instrumental values. Intrinsic values are values that are valuable for their own sake. Instrumental values are those which are, by contrast, pursued for the sake of another value. Typical examples of instrumental values are effectivity and efficiency. Both values assume that there is a goal, to which effectivity and efficiency are directed, so that they are not strived for for their own sake. A typical example of an intrinsic value is the good life. There are also values that do not fit so neatly into these two categories. Many important moral values that play a part in the designing of technical artefacts are indeed of such an intermediate nature. One might think for example of values such as safety, sustainability, health, privacy and justice. Health is something that might be seen as an instrumental value leading to a good life, but it is also, to a certain degree, a value in itself; it not only leads to a good life but is also itself part of a good life.

Norms are prescriptions for action. They are usually more concrete than values; they are, in fact, often directed at the realising of values. One might, for example, associate the value of traffic safety with a whole range of norms such as 'drive carefully' and 'give way to traffic from the right'. Norms can also be based on conventions. In the Netherlands, for instance, we have the convention that traffic coming from the right and not from the left has right of way on cross roads. Abiding by that convention is something that contributes to the realising of the value of road safety.

In their professional practice, engineers are inevitably involved with a vast number of norms. One need only think of all the technical norms that prescribe how an artefact must be designed. Another example is professional codes of conduct. In most countries, professional associations of engineers have formulated codes of conduct. This includes organisations like the National Society of Professional Engineers (NSPE), the American Society of Civil Engineers (ASCE) and the American Society of Mechanical Engineering (ASME) in the United States, the European Federation of National Engineering Associations (FEANI) in Europe, and the Dutch association of engineers KIVI-NIRIA in the Netherlands. Such codes of conduct include, for instance, stipulations that relate to the proper exercising of the profession such as the need to avoid or else report possible conflicts of interest, not yielding to bribery, keeping abreast of technological advancements, treating others and other cultures with respect, mentioning sources and maintaining open forms of communication and cooperation.

Alongside norms and values, also *virtues* can be distinguished. Virtues may be described as positive character traits such as courage, restraint and empathy. Virtues do not directly prescribe actions, but they are expressed in people's actions. A brave individual behaves differently from a coward. In addition to general virtues, professional virtues may be distinguished. These are those virtues that contribute to the proper exercising of one's profession. Examples of professional virtues, in the case of engineers, are competence, accuracy, creativity, perseverance and honesty.

Like norms, values and virtues, *responsibility* is an important concept in the area of ethics. If we hold someone responsible for something that happened, we usually mean that we expect that person to account for what happened. One can hold people accountable for actions or for the consequences of actions, such as when a bridge collapses. Responsibility is not only relevant in retrospect, i.e., if something went wrong, but also prospectively. If someone is responsible for something in a prospective sense, then that same person is expected to act in such a way that undesirable consequences are, as far as reasonably possible, prevented. An important question in engineering ethics is how far an engineer's responsibility extends. Generally, it is presumed that one can only be held responsible for matters that one has the power to influence. That is the reason why one cannot usually be held responsible for the behaviour of others or for natural disasters, which cannot be influenced by humans. In Chapter 7, we will take a closer look at the conditions under which a person can reasonably be held responsible. For the time being we shall mainly confine ourselves to the notion that responsibility depends on control or influence.

Nobody can deny that engineers have an influence on what they design even though the degree of influence is often curtailed by employers or clients. But what is the moral significance of what engineers design? In Section 1.5, we discussed the view that technology is a neutral instrument that only gains moral significance when used. This point of view considerably limits the responsibility of engineers. For by this view, engineers may still be responsible for designing artefacts that from a technical point of view function well, but possible negative consequences of technology related to its use are the responsibility of the user and not the engineer. As was argued in Section 1.5, the thesis that technology is merely a neutral instrument is problematic. The sphere of influence exercised by engineers extends beyond that of simply designing neutral instruments. This is a point that most engineers also recognise. For example, the preamble of the code of conduct of the NSPE states that:[15]

> *Engineering is an important and learned profession. As members of this profession, engineers are expected to exhibit the highest standards of honesty and integrity. Engineering has a direct and vital impact on the quality of life for all people. Accordingly, the services provided by engineers require honesty, impartiality, fairness, and equity, and must be dedicated to the protection of the public health, safety, and welfare. Engineers must perform under a standard of professional behavior that requires adherence to the highest principles of ethical conduct.*

Obviously, this does not mean to say that engineers can be held responsible for all the consequences of technology. We shall begin our exploration of their responsibility by elaborating on the idea presented in the previous chapter that engineers do not exclusively design artefacts but also use plans.

3.2   ENGINEERS AS DESIGNERS OF ACTIONS

If it is indeed true that engineers not only design artefacts but also use plans, then what are the implications for their responsibility? When engineers design use plans, they do not only come up with an artefact but also with a set of stipulated actions intended to achieve a certain goal. One

---

[15]NSPE. 2010. *NSPE Code of Ethics for Engineers*. National Society of Professional Engineers, USA 2007 [cited 10 September 2010]. Available from `http://www.nspe.org/Ethics/CodeofEthics/index.html`.

might assert that engineers therefore prescribe certain actions to the users and thereby design the norms for proper use of the artefact in question. Depending upon the exact nature of their content, these norms can also have moral relevance. The goal realised by the execution of a use plan is often morally relevant. A goal can be morally praiseworthy (for instance, if a road bump is introduced to enhance road safety), morally neutral (if one travels from *a* to *b*) but also morally reprehensible (if a person murders someone).

What are the conditions that a use plan has to satisfy if it is to be morally acceptable? A morally acceptable use plan will not at any rate be based on a morally reprehensible objective. That, though, is not the only requirement that we should pose for a use plan. Another requirement is that the execution of the use plan is actually conductive to attaining the stated goal or, in other words, that the plan is effective. The use plan must furthermore be realistic for the users. For example, a use plan that demands of users that they carry more weight than is physically possible is not a good use plan. Indeed, different users will invariably have different skills, strength and statures. The elderly, for instance, often lack the strength to use traditional tin openers, which means that they are excluded from using such artefacts and from benefiting from the advantages that they bring. Such exclusion does not have to be a moral issue if, for instance, alternative adapted tin openers are available at a reasonable price. The situation only becomes problematic when certain groups are excluded from vital social services. One might think, for example, of train ticket-dispensing machines that old people often find hard to operate. This would be especially morally problematic if there were no ticket purchasing alternatives for the elderly.

One might also wish to demand that a use plan can be executed and can realise its aim without producing any unacceptable side-effects. If electric sawing machine users run a substantial risk of sawing off their fingers when they use the device according to the use plan, the use plan in question cannot be termed a good one. Finally, the use plan also has to be properly communicated to the users. This can be done via a manual but also by means of instructions printed on the appliance itself or, for more complicated appliances, through special courses. The last two requirements are described in the professional code of conduct of the KIVI-NIRIA, which stipulates that[16]:

> *(1.3.2) The engineer must provide manuals (stipulating the relevant standards and quality norms) so that the user has the opportunity to make safe use of the products and systems for which the engineer is responsible.*

Imagine that an engineer designs a technical artefact and provides a use plan that conforms to all the above-mentioned requirements. If something then still goes wrong with the artefact, or if unintended and undesirable side-effects materialise, is that then the responsibility of the user rather than the designer? There seems to be some sense in this. If users deviate from the use plan, surely, it is they who should be blamed and not the designer, certainly if the plan was feasible and met all the other listed requirements. If a person drives too fast in his car and causes an accident, then we are inclined to blame that person and not the designer of the car in question. Clearly, the designer cannot control

---

[16]http://www.kiviniria.net/CM/PAG000002804/Gedragscode_2006.html, last visited on 23rd June 2008. Our translation.

the behaviour of users, and therefore designers can hardly be held responsible for the behaviour of drivers. However, matters change if the undesirable consequences can be traced back to a deficiency in the use plan or to a deficient communication of the use plan.

Even if there is nothing wrong with the use plan, one may still question whether the responsibility of the designer always ends with the creation of a good use plan. Let us take as an example the development of automatic weapons. Those are firearms that are capable of discharging large rounds of ammunition in a short period of time. Such shoulder arms, like the AK-47 and its successors, better known colloquially as the *kalashnikov*, are currently very popular weapons in, for instance, civil wars or guerrilla conflict situations where they are often wielded by child soldiers. To what extent may the developers of such firearms, such as General Kalashnikov, be held partly responsible for the way in which such guns are used and for the many fatalities they cause? According to the analysis given above, their degree of responsibility is closely allied to the exact use plans provided for such weapons. Just imagine that the use plan communicated confines itself to the use of such a weapon for purely professional military ends, for the purposes of proportional violence, that is to say, violence that is proportional to that which it endeavours to combat, for instance, because it is necessary to reach legitimate military targets or in order to guarantee security. Does such a use plan release the designers from responsibility? One good reason for doubting this is because studies conducted since the Second World War have proved that machine guns considerably lower the violence threshold. When eye-to-eye with the enemy, soldiers are much sooner inclined to shoot with a machine-gun than with a traditional gun. Apart from anything else, largely thanks to the number of bullets that can be dispensed in a short space of time, automatic firearms are much more effective when it comes to killing the enemy. More so than with traditional guns, these weapons therefore encourage excessive acts of violence. Even if this was not part of the original use plan, it seems reasonable to assert that the designers are partly responsible for the consequences, especially if one thinks that such potential effects could be known from existing studies. Regarding the use of such weapons in civil war situations and by child soldiers, the matter is somewhat more complex. The weapons possess certain features that make them ideal for such use – they are relatively light and easy to operate compared with other weapons and require little maintenance – but one may question to what degree this type of use could reasonably have been foreseen at the time when the weapons were developed. Nevertheless, it can be concluded that the responsibility of engineers sometimes extends further than the use plan.

The argumentation above indicates that we need a term that denotes various forms of use that may not literally be seen as part of the use plan but which the technical artefact in question does invite. A term that Madeleine Akrich and Bruno Latour, two sociologists of technology, have come up with is '*script*'.[17] A script may be described as the behavioural norms which – intentionally or not – are built into a certain technical artefact. A nice example of this is the automatic seatbelt. In the case of automatic seatbelts, the car simply will not start if the seatbelt is not used. That therefore

---

[17] See, for example, Akrich, M. [1992], Latour, B. [1992] and Verbeek, P. [2005].

forces the driver to fasten his seatbelt. The inbuilt norm is thus 'do not drive your car without a seatbelt.' With the help of this norm, the value 'safety for the car driver' is promoted.

Integrating behavioural norms into technical artefacts has undoubted advantages. By using automatic seatbelts, traffic safety is enhanced, the use of turnkeys ensures that all metro passengers pay for their tickets and by operating a prepaid system for domestic energy consumption people are able to control their spending on gas and electricity more stringently. In all these examples technology is implemented for the regulation of human behaviour. There is, however, a fundamental ethical question raised by such examples: what gives designers the right to regulate the (moral) behaviour of users via the products that they design?

An important other way in which human behaviour is regulated is by means of legislation. However, legislation has to be first approved by democratically elected bodies, and this gives them a certain legitimacy. It would seem that such legitimacy is lacking in the case of designers if they regulate the behaviour of others through their designs. It would thus seem that this creates additional responsibilities for them to shoulder.

It is indeed important to realise that the degree to which artefacts dictate or even enforce certain kinds of behaviour is partly dependent on design choices. There are, for instance, different car seatbelt designs which impose their use to varying degrees. The traditional kind of seatbelt does not force people to use it, but there are also car systems that give some kind of a warning if the driver is not belted up. In the latter case, the driver is not forced to use the seatbelt, but he or she is encouraged to do that.

3.3   VALUES AND THE DESIGN PROCESS

We have seen how in the designing of technical artefacts, certain behavioural norms for users are simultaneously designed. Those norms sometimes contribute to the realisation of certain values, such as traffic safety in the case of automatic car seatbelts. In addition to that, technical artefacts can also contribute to certain values in negative and positive ways through their side-effects. A car not only transports the user from a to b, but it also produces exhaust containing polluting substances that contribute negatively to values such as human health and the environment. The extent to which a technical artefact positively or negatively contributes to certain values does not thus depend solely upon how it is used but also on how it is designed. Values therefore play a part in the design process in different ways.

For example, values such as sustainability and safety may have a role to play in the formulation of design requirements. One need only think of the design requirement that a certain appliance may not use more than a stipulated amount of energy or that the chance of a certain component breaking down should be below a certain probability. At the same time, values can be translated into design requirements in different ways. For instance, the matter of whether the value of safety, in the case of vehicles, is translated into safety for passengers or safety for other road users, such as cyclists and pedestrians, makes quite a difference. Cars that are safe for their passengers will not necessarily also be safe for other road users.

Apart from being translated into design requirements, values also have a part to play in the choices that are made between different concept designs. It is frequently the case that during a design process, different options are compared with the aim to select ultimately the 'best' alternative. Values can inform such decision-making. Different potential choices can be compared from the point of view of the extent to which they realise certain values. Often, as part of the process, the values are first translated into different criteria.

In many cases, not all the relevant values can be simultaneously realised, or at least not optimally, during the designing of certain technical artefacts. The safest car is not necessarily also the most sustainable or the cheapest one. Such kinds of choices, where the values at stake conflict, are often hard to make. The question as to how, in such cases, the best decisions can be made is not merely a practical matter but also a philosophical question. And the answer to that question also depends on the answer to the philosophical question as to what, exactly, values are. Some philosophers maintain that there is only one ultimate value and that all other values are only of significance insofar as they contribute to that one supreme value. Adherents to the theory of utilitarianism, such as Jeremy Bentham, maintain that human happiness or pleasure amounts to the ultimate value and that all else can be measured in terms of the extent to which it contributes to that one value. The cost-benefit method that is quite often used in engineering practice is motivated by similar ideas. One can, though, seriously question whether human happiness really is the one and only true value. Focussing on well-being ignores the distribution of burdens and benefits; often a fair or just distribution of burdens and benefits is seen as an important value that is valuable independent from maximising individual well-being.

Yet other philosophers, like Immanuel Kant, maintain that all value conflicts can in the end be resolved through reasoning and by appealing to human ratio or 'good will'. If that is the approach chosen when conflict arises between values, it is important to first establish why a value is worth pursuing. Speed bumps, for instance, increase the safety of playing children but diminish the freedom of car drivers. Both safety and freedom are values. In this particular case, there would appear to be good moral grounds for placing the freedom of playing children above the freedom of car drivers. Generally speaking, freedom is seen as something worth pursuing whenever it gives people the chance to give meaning to their own life and to make their own moral choices. Freedom is not about being totally free to do as one pleases or to curtail the freedom of others for no good reason. Indeed, one may query whether having the freedom to drive as fast as one wants in built up areas can truly be termed a kind of freedom that is a value worth aspiring to. In this case, it seems, at any rate, morally defendable to curtail the freedom of car drivers in the interests of playing children and other road users. Sometimes value conflicts can be resolved through reasoning, but one has to question whether this is always the case.

A third way of coping with values during design processes is by adopting a *value sensitive design* approach. The idea behind this approach is that at the start of a design process one establishes which values are morally of importance when it comes to designing a technical artefact. Subsequently to the content of these values, one establishes why the values are of moral significance and then they are translated into certain design requirements. These values can also come to play a part when choosing between various alternatives. When engineers design in a value sensitive way, they seek technical solutions that help them to realise as well as possible all the relevant values. Values that may conflict in one particular design do not need to conflict in another innovative design. Values are thus able to steer the design process possibly leading to socially better technologies.

3.4   THE DEVELOPMENT OF QUIETER AEROPLANE ENGINES

In this section, we shall illustrate the part played by values in the design process by a discussion of the designing of aeroplane engines.[18] These kinds of engines have to satisfy a vast number of design requirements. Some of the design requirements have to do with the reliability and safety of the engines. The moral value that is of relevance here is safety, both for passengers and people on the ground. Aeroplane engines also have to be capable of fulfilling certain performance demands like, for instance, a certain thrust or speed. Here the underlying moral value is human well-being; the underlying conviction being that the ability to enjoy air travel (and the faster, the better) is something that serves human well-being. One can, of course, debate whether this is really the case. Aircraft engines also have to comply with certain environmental norms with respect to, for example, fuel consumption or the emission of various substances. The underlying value 'environment' can be justified in terms of the moral obligations that we have towards future generations but also in terms of the intrinsic value of nature. Finally, aeroplane engines also have to observe various noise level requirements; again, the value at stake is human well-being, both for passengers and people on the ground.

It is definitely not easy to simultaneously respect all these different values when developing aeroplane engines. With the advent of jet propulsion in the post World War II era, it became possible to fly at much higher speeds, but the new jet engines consumed much more fuel than the old propeller-driven aircraft. The use of jet propulsion also led to much more noise hindrance on the ground, though it must be admitted that for passengers jet engines were, and are, much more silent than traditional propeller engines. The introduction of jet propulsion therefore gives rise to a number of moral questions. One question one might ask is whether the speed gain derived from jet power can justify the higher consumption of fuel. The values of human well-being and the environment seem to clash when one poses this question in this way. The dilemma may be formulated in terms of the well-being of present generations against that of future generations or in terms of the well-being of people who fly and those who do not. The last two formulations foreground the fair distribution of burdens and benefits. Another moral question linked to the jet engine is to just what extent the noise hindrance for plane passengers weighs up against hindrance for those living in the vicinity of airports; the jet engine is quieter than the propeller engine for passengers but noisier for people on the ground.

[18]Van de Poel, I. [1998].

Aircraft engine development therefore raises a whole host of moral questions. In a number of instances, there is evidence of conflicting moral values in the design process. As we have seen, in some cases, such conflict can be formulated in terms of fairness, namely in terms of the fair distribution of burdens and benefits, for example, between passengers and people on the ground or between present and future generations. It would seem difficult to argue that people on the ground should be subjected to greater noise hindrance just so that those who choose to fly can do so in greater comfort, with less noise and in shorter journey times. In the years since the jet engine has become a fact of life, most governments have laid down and gradually tightened up noise limit requirements for the aviation sector. In this particular instance governments chose to implement legislation that would safeguard that certain moral values – concerning the well-being of people on the ground – would be minimally met.

Another way in which conflict between the different moral values in aircraft engine design is tackled is via the route of technological innovation which has, in fact, made feasible to realise, to a large extent, divergent values simultaneously. An important incentive in the innovation process was the aim to develop an engine that was more efficient than the first generation of jet engines but, at the same time, faster than the propeller-driven engine. To properly grasp this development, it is necessary to briefly sketch several relevant features of jet engines.

With traditional propeller propulsion, the engine power is converted into thrust by means of a propeller that gives a large volume of air a small speed acceleration. With the jet engines that were developed just after the Second World War, the thrust was obtained by producing a jet flow at relatively high speed. In that way, a small volume of air was given a high velocity increase. The operational principles of jet power are thus essentially different to propeller power. Hence, the reason that in Section 2.2, mention was made of radical designing.

The jet engine and the propeller engine represent two different kinds of trade-offs between efficiency and speed. The first jet engines were very fast but not very efficient. By contrast, the propeller engines are efficient but not very fast. At a certain flying speed, an aeroplane engine – if one ignores mechanical losses – is most efficient when the velocity of the outflowing air (or in the case of propeller engines the compressed air) is almost equal to the cruising speed. In view of the fact that most jet engine civil aircraft usually flew at speeds lower than the outflow speeds of the early jet engines, there was room for considerable efficiency gains if ways could be found of lowering the velocity of the air emitted from the engine. This required further technological innovation, eventually resulting in the development of what is called the bypass-engine or turbofan.

A turbofan is a jet engine in which part of the air bypasses the combustion chamber and is accelerated by one or more fans (see Figure 3.1). The bypassing air is then mixed with the (faster) burned fuel giving the device its net thrust. The total amount of air that flows through a turbofan is considerably more than the amount that flows through a comparable traditional jet engine. The result is that the average jet speed is lower, which makes the turbofan more efficient but at the cost of lower maximum speeds.

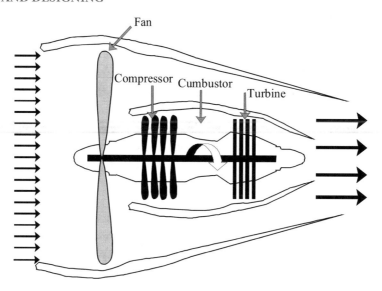

Figure 3.1: The turbofan: Part of the air by-passes the combustion chamber. The turbine drives one or more fans. These fans accelerate the bypassing air, which mixes at the back with the (faster) burned fuel. The net thrust results from the speed of the exhaust air.

Turbofans are not only more efficient, but they also offer good opportunities for noise reduction. With traditional jet engines, the main reason why they create so much noise is because of the speed difference between the jet stream and the air cruising speed. When the turbofan was invented, this noise factor was considerably reduced. On the other hand, the turbofan has more mechanical parts, such as an extra fan, which also generates noise. With the aid of knowledge gained from the field of aero-acoustics, a field born of the need to reduce aircraft engine noise, it was discovered that with the aid of design adjustments, the extra sources of noise could be considerably reduced so that aircraft engines could be made quieter.

The evolution of aeroplane engines might be viewed as a type of value sensitive design. Indeed, through technological innovations, it became possible to more easily satisfy a number of partly conflicting values. The engineers and scientists involved did not consciously or explicitly reflect on the moral values at stake during the design process. Nevertheless, the whole evolution of aeroplane engines underlines the fact that certain moral values are sometimes best achieved through technological innovation.

Through technological innovation planes have become gradually less noisy over the course of time. This has not, however, solved the problem of noise hindrance on the ground. An important reason for this is the increase in the number of aeroplane movements. Up to a point, that increase has been facilitated by the fact that modern planes are quieter. In this respect, the effort to resolve the noise problem through innovation may be seen as a '*technological fix.*' The inherent danger of such a *technological fix* is that engineers try to use technological means to resolve problems that have social origins. In this particular case, the increasing desire to fly was not acknowledged, which meant that the problem was not really resolved.

3.5   REGULATIVE FRAMEWORKS[19]

So far, we have discussed norms and values that engineers help to realise by means of their designs. We have seen that the topic leads to a number of ethical questions: What side-effects are acceptable? Do engineers have the right to regulate or try to regulate the behaviour of users? And how must different values in the design process be weighed up against each other? When it comes to the matter of answering these questions, engineers are fortunately not usually completely on their own. In many cases certain norms. and rules have already been established over the course of time on ways of dealing with these issues. As far as most artefacts are concerned, there are usually laws and regulations that have been promulgated by governments. Furthermore, for many technical artefacts, there are often technical codes and standards that can be quite detailed [Hunter, T., 1997]. Technical codes are legal requirements that are enforced by a governmental body to protect safety, health and other relevant values. Technical standards are usually recommendations, rather than legal requirements, that are written by engineering experts in standardisation committees. Some artefacts have to be certified before they can be used. Aeroplanes, for example, are only allowed to fly if they have been certified. One of the aspects covered by certification is airworthiness; another is noise reduction.

Although some of the rules mentioned apply to all engineering areas, a great many are product-specific or dependent on the type of artefact developed. All rules that apply to a specific kind of technical artefacts form together what may be called a *regulative framework*. We briefly discuss some of the main elements of the regulative framework for pressure vessels. In the European Union (EU), the main legal framework is the EU Pressure Equipment Directive (Directive 92/23/EC). Member countries of the EU are required to comply with this framework, but they may stipulate additional demands that have to be met by pressure vessels in, for instance, the area of safety. In the Netherlands, requirements are, for example, laid down in the Nuisance Act (*Hinderwet*), Law on goods (*Warenwet*), Pressure Vessel Decree (*Drukvatenbesluit*), and the Pressure Systems Decree (*Drukapparatuurbesluit*). Such legal regulations do usually not specify in detail how exactly their requirements must be met. Technical standards are usually much more detailed; they often lay down various construction details or calculations. They may, for example, prescribe that a certain type of pressure vessel must have a given wall thickness. In some countries, following technical standards is legally obliged; in other countries, following technical standards is taken as evidence that the relevant legislation and regulations have been satisfied, although other ways of meeting the legal requirements may be possible and allowed. Standards may be internal to a company, to a consortium of companies or industry-wide. Industry-wide standards are usually formulated through national standards institutes, like the American National Standards Institute (ANSI) and the International Organization for

[19]Van Gorp, A. [2005].

Standardization (ISO). ANSI has accredited a number of organisations as standards, developing organizations, like the American Society of Mechanical Engineers (ASME). These organisations organise the process of standard formulation, which involves the relevant stakeholders and which has to meet the requirements formulated by ANSI to guarantee openness, transparency, balance of interest and due process. European standards are formulated through the CEN, the European Committee for Standardization. The procedure is comparable to that of the ANSI.

A regulative framework may be seen as a kind of reified morality. It lays down prescriptions for right action in rules. Such rules are usually based on past experience, like certain kinds of accidents that have occurred with a given type of technology, or the discovery of certain negative side-effects. Often a regulative framework is established in the wake of public unrest or public debate. In the case of legislation, the rules laid down result from democratic decision-making.

A regulative framework can support engineers when confronted with ethical questions in their professional practice. Insofar as the framework is democratically legitimised, it also provides a solution for the previously mentioned matter of when and whether engineers are allowed to regulate the moral behaviour of users and others. Nevertheless, one cannot blindly follow a regulative framework. Even if the regulative framework amounts, in effect, to reified morality, that does not mean that it is always or necessarily morally desirable or even morally acceptable. If, as an engineer, you want to apply a regulative framework, a number of conditions have to be satisfied. One can, for instance, think of the following series of conditions:[21]

- The framework must be *complete*, which means to say, that it must provide an answer to the question of how to act in specific situations without leaving relevant aspects of the situation out of the picture.

- The framework must *not contain* any *contradictions* or inconsistencies.

- The framework must *not be ambiguous*, that means to say, there must be consensus on how the framework applies in a specific situation.

- The framework must be *morally acceptable*.

- The framework must be *observed* and not just paid lip service.

Whether a regulative framework indeed exists that does fulfil the stipulated conditions partly depends on the kind of design process. Empirical research suggests that regulative frameworks are more common in the case of normal designing than with radical designing. This is understandable. A number of the rules of regulative frameworks are founded on the operational principle and the normal configuration of an artefact. This applies especially to more detailed rules. Very general rules – like, for instance, that safety must be taken into consideration – do not depend on the operational principle or the normal configuration of products but more specific rules that guarantee safety –

---

[21]The conditions given below derive from Grunwald, A. [2001]. In certain places, the formulation has been adapted. In the fourth condition, Grunwald states that all those involved must accept the framework.

like, for instance, that the walls of a pressure vessel must have a certain thickness – do depend on such principles or configurations. The dictated thickness of a pressure vessel depends, amongst other matters, on the chosen type of material and the standard configuration of an installation. The rules concerning the wall thickness of steel barrels cannot simply be applied to pressure vessels made of synthetic materials. Sometimes the requirements can even be nonsensical because they apply to quantities or measurement methods that do not properly relate to the new situations or are simply no longer applicable. For such reasons, regulative frameworks may no longer apply in cases of radical design.

Even though with normal design there is usually a regulative framework, it does not always satisfy the above-mentioned conditions. One could take as a good example crash tests for cars. These tests place more emphasis on the safety of the car occupants than on the safety of other road users such as pedestrians and cyclists. A car that is found to be safe for its occupants will not also necessarily be safe for other road users. Some of the measures taken to increase the safety of passengers, such as increasing the weight and rigidity of the car, might even reduce the safety of cyclists and pedestrians. On moral grounds, it could be argued that the safety of cyclists and pedestrians is more important than that of people in the car. After all, those who get into a car take the risks attached to car transport more voluntarily than those who are not in the car and who either cannot or who can hardly be said to choose to take such risks. Apart from anything else, people who are not in a car do not have any of the benefits of car transport while the occupants do, unless one presumes that people not in the car travel by car roughly as frequently as those in the car, but one can question whether that presumption is applicable. In such cases, the decision to either accept or not accept the regulative framework is therefore a moral choice that engineers have to make.

All in all, there are three conceivable situations as far as the availability of a regulative framework is concerned. In the first place, there might be a framework available that fulfils all the mentioned conditions. In these kinds of situations, engineers can be guided by this framework when making their ethical choices during the design process. In the second place, there may not be a framework that is applicable to the relevant design. This kind of a situation is quite thinkable in radical designing. In such cases, the engineers have to make their ethical choices either individually or in collaboration with others such as the users, regulatory bodies or other stakeholders. In this situation, the engineers have more own responsibility for the artefact they design because they are less able to rely on already socially sanctioned rules. In the third place, it might be possible that a framework does exist but that it does not satisfy one or more of the mentioned conditions. Precisely what is the most sensible strategy in such a case will partly depend on the condition that has not been met. We shall confine ourselves here below to the question of what engineers can do if the existing framework is morally unacceptable or morally disputable.

As an engineer, there are different courses of action one can take if a regulative framework is not acceptable. To start off with, you can try to alter the framework. Engineers may be involved in committees that draw up technical codes and standards or think up test procedures, and they are thus able to have an influence on these aspects of the regulative framework. In the second place, complaints can be lodged with bodies that are able to influence other aspects of the framework like, for instance, the government, the parliament or a professional organisation. In the third place, one can choose to deviate in the design process from certain elements of the regulative framework. Existing frameworks often provide a certain margin of flexibility for such deviation, but with legal rules, the engineer does, of course, have less room for manoeuvre. Finally, you can opt for radical designing where the existing framework will not be applicable.

Though there can be great advantages attached to radical designing there are also often new kinds of risks that arise. One such example is the introduction of the jet engine discussed above. Not very long after the jet engine had been introduced to the world of civil aviation, two such aeroplanes – with the rather unfortunate name Havilland Comet – crashed. The problem did not so much reside in the engines themselves as in the fact that jet-powered planes flew at much higher altitudes than previous aircraft had done. This meant that cabins had to be pressurised to make flying comfortable for the passengers. As a result, some points of the fuselage were subjected to greater stresses than previously. That, in turn, led to metal fatigue and ultimately to disaster. What this proves is that existing frameworks designed to safeguard certain moral values – in this case safety –cannot be automatically transferred to radical designs. When choosing between normal and radical designs there are also, therefore, moral considerations alongside the technical considerations. Such moral considerations include the moral acceptability of the existing regulative framework, the degree to which a radical design can probably better satisfy important moral values and the level of risk that is introduced with radical designing.

3.6   CONCLUSION

Engineers change the world for the better and for the worse by means of the artefacts they design. One way in which they do that is by designing behavioural norms for users. Those behavioural norms can be part and parcel of the use plan for an artefact that is communicated to the users, but they can also be inherently present in the artefact's design. We have seen that a morally acceptable use plan must conform to a number of conditions: it must serve a morally acceptable goal, it must be effective, possible to execute, have no unacceptable side-effects in practice and it must be properly communicated to the user. Through the artefacts that they design, engineers also influence the realisation of certain values. In the design process, such values are often translated into design requirements; moreover, they are often guiding in choosing between different concept designs. Various values can conflict, but through value sensitive innovation strategies, engineers can endeavour to temper the effects of such conflicts. Finally, we considered the matter of the extent to which engineers are able to fall back on more or less generally accepted norms and rules when making their choices during the design process. It emerged that this is usually an option in the case of normal designing but not with radical designing. In the latter case, engineers have to make more ethical choice decisions themselves, which means that they also bear greater moral responsibility.

In Chapters 5 and 6, we shall expand our analysis of technological matters to include the social and societal questions surrounding technology. In Chapter 7, we shall then return to the matter of the ethics of design and the responsibility of engineers for the unintended (social and societal) consequences of technology.

3.7   A FEW MORE ISSUES

In the present chapter, we have examined the conditions with which a morally acceptable use plan must comply. We also introduced the notion that the responsibility of the designer can sometimes extend beyond the designing of a good use plan; the designer can sometimes even be held partly responsible for types of use that the technical artefact makes possible or invites. What, in this connection, can one say about the responsibility for misuse of a technical artefact and the ensuing consequences? In such cases, is the user responsible or fully responsible; does the designer also have some responsibility and, if so, when? When answering these questions, it is useful to indicate what precisely you perceive to be misuse: does it relate to all kinds of use not included in the use plan? Is it all kinds of use for which the artefact is not suitable? Or still something else?

A second, more difficult question is whether and, if so, under what conditions engineers may try to morally lay down the law for users through their designs. This is sometimes viewed as a type of paternalism with engineers, wrongly presuming that they know better what is good for people than people themselves. Perhaps, in this area, regulative frameworks hold the answer: they are, at least, generally democratically legitimised. But what can one do if there is no regulative framework or if the framework that is available is morally unacceptable? Are engineers then allowed to lay down moral laws for others?

CHAPTER 4

Technological Knowledge

In this chapter, we return to engineering practice; to discover what forms of knowledge are relevant in that area. The idea underscoring much philosophical work on this subject is that technology is nothing other than applied science. We shall demonstrate that this idea is wrong: engineers do more than simply use scientific or applied scientific knowledge, they *develop* own forms of knowledge. We will give a number of examples of such specific *technological* knowledge and present a number of relevant specific features and characteristics.

4.1   ENGINEERS AND KNOWLEDGE

In the previous chapters, we stressed how important knowledge is for engineers. In Chapter 2, we described what kind of knowledge is needed to be able to design a suitable artefact: one has to know what the user wants, what is feasible, how artefacts are generally used, what other artefacts are used, how materials behave, et cetera. Lack of such knowledge is one of the many reasons why a design can be doomed to failure.

Engineers therefore need to have knowledge and to know how to apply it, but it would seem that accumulating new knowledge is not their main objective. Sometimes, this has led to the conclusion that engineers do nothing other than find practical applications for knowledge amassed by others. This is the image of technology as applied science: scientists gather knowledge and create theories, and engineers apply that knowledge in order to design artefacts. The result of all that effort is often practically useful or even indispensable, but it does not lead to new knowledge about the world. We label this the 'applied-science-view'.

A particularly arresting description of that view is encapsulated in the slogan for the '*Century of Progress*' exhibition staged in Chicago in 1933: '*Science Finds – Industry Applies – Society Conforms*'. This conveys the image of a perpetual flow of knowledge trickling down from the realms of science to industry, where the engineers create products and society adapts its behaviour to those artefacts. The slogan is essentially an ideal that is presented as a description: industry need not consider the demands and behaviour of users, just as scientists must be free to develop knowledge without burdening themselves with potential applications. Engineers are supposed to be consumers, not producers of knowledge, just as all other members of society are supposed to be consumers, not producers of new devices.

The applied-science-view cannot be separated from the way in which engineers perceived themselves. From the dawn of the Industrial Revolution until far into the last century, traditional engineering disciplines such as civil and mechanical engineering shifted increasingly from continuations of practical and traditional craftsmanship to the scientific end of the spectrum. Even the relevant education was, in the process, reformed. Instead of providing practical, profession-oriented training, engineering curricula were organised in such a way that students were taught, above all else, just how to apply the theories gained from applied scientific research. The underlying notion was that academically trained engineers were not craftsmen but scientists, and that they should thus be formed as much as possible in the mould of such academics.

After about 1960, a clear reaction to this school of thought could be detected. Increasingly fewer people, engineers and others, believed that engineering disciplines and practice could be or even needed to be reshaped, according to some kind of scientific model. Herbert Simon influentially pointed out that engineering disciplines are 'sciences of the artificial', just like for instance computer science, and that they differ inherently from the natural sciences.[21] According to Simon, the core competence of natural scientists is their ability to understand, describe and explain reality whereas scientists of the artificial were in the business of changing the world for practical purposes. An even stronger reaction to the 'scientification' of engineering was voiced by Donald Schön.[22] Schön argued that the emphasis on formalisation and explicit procedures – also by Simon – creates a false impression of the work of many professionals, including engineers. He stressed the importance of personal experience and training, especially in the design process, because crucial skills could only in that way be learned. To a large extent, Schön therefore wanted engineering to go back to its craftsmanship roots.

After 1970, the applied-science-view had been abandoned by most authors who reflected on the nature and methods of engineering. One important counterargument to the view was that there had been all kinds of technological developments which only much later – if at all – could be scientifically substantiated. The first generations of steam engines were, for instance, built without scientists even knowing how to describe how those particular artefacts worked. Similarly, transistors and, more recently, high-temperature superconductors were created before scientists were able to precisely comprehend what physical processes were responsible for the successful operation of such innovations. In these cases, revolutionary industrial applications came long before scientific understanding.

For this reason, it is impossible to defend that engineers always apply what was first theoretically mapped out by scientists. That does not mean that science is not a good or even optimal basis for technological developments. Very recent developments, for example in the fields of nano- and biotechnology, cannot be understood in isolation from fundamental scientific knowledge; and the curricula of most engineering courses, especially at technological universities, are to a large extent made up of core scientific subjects such as mechanics and thermodynamics.

Moreover, it may be an exaggeration to say that engineers only apply scientific knowledge, but that does not imply that engineers actually develop knowledge themselves. There are, nevertheless, enough reasons to presume that there is such a thing as 'technological knowledge', that this form of

---

[21] Simon, H. [1967].
[22] Schön, D. [1983].

knowledge differs in certain respects from scientific knowledge and that engineers make an important contribution to the development of technological knowledge. In the next section, we will provide two examples of technological knowledge that has come to play a part in different engineering disciplines. In Section 4.3, we shall then go on to describe a number of important features of technological knowledge.

4.2   WHAT ENGINEERS KNOW: TWO EXAMPLES

One way in which the applied-science-view can be successfully undermined is by showing in detail how artefacts are designed and technically developed, and by explaining which kinds of knowledge are required in the process. Some classical case studies of this type have been conducted by Walter Vincenti. His book *What Engineers Know and How They Know It* has formed the basis for many studies, philosophical and otherwise, of technological knowledge since 1990.

Vincenti describes various types of engineering knowledge, especially knowledge which is used and developed during the design process. In Section 2.2, we already referred to this in our characterisation of technical designing. Two particular examples described by Vincenti show how broad and varied the knowledge of engineers is: in both examples, a different type of knowledge is used and developed – the first fairly theoretical, the second more practical – and both types are different from scientific knowledge as well.

The first example is a calculation method, 'control-volume analysis', that is used widely by engineering scientists and designers. In this method, any spatially well-defined volume may be chosen, through which fluid or heat flows, or work is transmitted. Applying the laws of physics to this volume and its surface then yields integral equations for both internal changes in the volume and the transport of liquids, heat or work across its boundaries (see Figure 4.1).

Since the equations are derived more or less directly from the laws of physics, control-volume analysis seems just a useful instrument for engineers who wish to keep their physical bookkeeping in good order – the method generates no new knowledge about the physical world.

Ultimately, though, this method involves much more than handy bookkeeping. The choice of the control-volume is of crucial importance for engineering purposes, although it is completely arbitrary from the perspective of physics. Vincenti, W. [1990, pp. 113; 115] quotes from the Reynolds and Perkins textbook on thermodynamics, which advises engineers to put the boundaries of the volume 'either where you know something or where you want to know something'. In other words, the method produces knowledge: no knowledge that lies essentially beyond the bounds of physics but certainly knowledge that is of direct relevance to engineers and irrelevant to physicists. The importance of that knowledge is revealed in actual applications of the method, in which the volumes usually follows the contours of various artefacts or components, such as steam engines, propellers, turbines and tubes. By choosing the control-volumes in this way, engineers are able to keep technically relevant records of, for instance, fluid flows in and out of a tube system, without having to worry about all kinds of internal physical factors like turbulence. Control-volume analysis therefore operates like a filter for useful, technological knowledge.

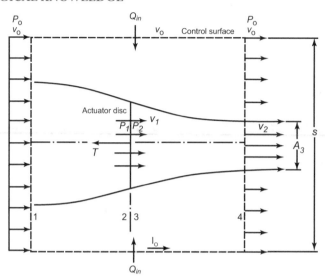

Figure 4.1: The control-volume analysis of a propeller [McCormick, B., 1979, p. 343].

The method is furthermore unique to engineering science. It is presented, together with all kinds of applications, in virtually all courses in and textbooks on engineering thermodynamics. But whereas students in applied physics or chemical technology are supposed to master the technique, it is unknown to most students in physics or chemistry. Vincenti gives the example of Zemansky's classical *Thermodynamics*. The only edition of that work that contains a description of control-volume analysis was especially written for engineers and co-authored by an engineer; there is no trace of the technique in other editions.

A second form of knowledge that Vincenti describes at length, reveals the importance of technical specifications in the design process. How, as a designer, do you arrive at these specifications? If you want to design a car that performs just as well as a current model but consumes 10% less fuel, the description of the specifications is relatively simple – even though they will undoubtedly present you with some difficult trade-offs later in the design process. But not every design process can be so exactly described; sometimes, it is a major problem to draw up the specifications in the first place.

Vincenti illustrates this with an example from aviation. In the early days of aeronautics, the primary concerns were direct performance features such as lift, speed and maximum altitude. After the 1920s, however, the attention of designers shifted to other, more qualitative aspects, such as stability and manoeuvrability. It was no easy feat to formulate specifications for those characteristics. Vincenti describes, in detail, how engineers searched for suitable variables. The task was to convert subjective perceptions of pilots, notably that of having control over the plane, into concrete specifications. Engineers did that in cooperation with test pilots: by ensuring that certain variables could be influenced by the pilot, the engineers could check whether the feeling of being in control had increased. Around 1940, a large number of relevant quantities had been defined on the basis of that knowledge. One of these, 'stick-force per g', was to become the most important specification. Vincenti adds that nowadays aeroengineers find it hard to imagine that other specifications were ever used, and that it did, in fact, take years to recognise the importance of the particular quantity they use. That proves just how important it is to have comprehensive and precise specifications: once they have been established, they are as 'natural' and 'self-evident' for engineers as Maxwell's laws are for physicists. In addition, the example demonstrates that identifying a new specification amounts to developing new knowledge. Such knowledge cannot be simply deduced from physical knowledge about the designed artefact, and it is indispensable to successful designing. Even if aeroengineers in the early 1930s would have described their plane design in all possible physical quantities, they would not have realised that 'stick-force per g' was the most relevant factor for designing a manoeuvrable aeroplane. One might even go as far as to posit that 'stick-force per g' is not a *physical* quantity but rather a *technical* one. If one views an aeroplane as a physical object, like a stone, then that factor has no special physical significance whatsoever. It only becomes of interest when one regards an aeroplane as an artefact that serves a certain goal and that people want to use it in a certain way.

4.3   FORMS AND FEATURES OF TECHNOLOGICAL KNOWLEDGE

Compared to scientific knowledge, technological knowledge has not received a lot of attention from philosophers. Nonetheless, it has a number of distinctive and interesting features. We shall discuss just four of them here. In so doing, we shall systematically examine one or more specific forms of technological knowledge that show these features. In this way, we demonstrate not just the uniqueness but also the diversity and scope of technological knowledge.

a. Technological knowledge is artefact-oriented.

Knowledge is often specified on the basis of the subject or, more precisely, the *domain* to which it pertains. Many descriptions of scientific areas are similarly domain oriented. Physics, for instance, is described as (the system of) knowledge of matter and energy, and economics as knowledge of the production, distribution and consumption of goods and services.

An obvious way to characterise technological knowledge would therefore be to specify its domain. Technological knowledge could, for instance, be said to be all about artefacts, technical or otherwise, or about objects designed by engineers for practical purposes (see Chapters 1 and 2). Some technological knowledge might be confined to one aspect of an artefact whilst other knowledge will relate to a whole collection of artefacts. Control-volume analysis can essentially be applied to any technical system that contains fluids. Knowledge about the navigability of aeroplanes was, initially, only related to the one model type on which the tests were carried out. Later, especially through defining a relevant quantity, it was extended to other types of aircraft.

Two kinds of technological knowledge that clearly concern artefacts are knowledge of *operational principles* and knowledge of the standard or *normal configurations* of artefacts. Knowledge of an operational principle is, simply put, knowledge of the way in which a particular thing works, that is to say, how it fulfils its function. One operational principle of an aeroplane may, for instance, be described as creating upward force or 'lift' by moving through air. A good knowledge of operational principles is indispensable to engineering practice and is always oriented towards a certain type of artefact. Much the same goes for knowledge of normal configurations, that is to say, the organisation of components in a given artefact with a certain operational principle. Anyone who sets about designing an aeroplane will base his or her ideas on existing designs by thinking, for instance, in terms of a structure that has two engines attached to the wings. This is not to say that an engineering designer can not in any way deviate from existing patterns, but he will have a repertoire of successful configurations. In combination, operational principles and normal configurations dictate the normal design context, the way in which a certain problem is typically resolved. Such solutions have often been able to prove their merits in practice and, therefore, also reflect designers' and users' experience.

b. Technological knowledge is directed at usefulness not at truth.

You cannot specify technological knowledge just in terms of its domain, although it may certainly be helpful. Most artefacts are also physical objects, so part of our knowledge about them is not exclusively technological but also scientific. Engineers are not interested in all this scientific knowledge. This, for instance, becomes clear in their use of control-volume analysis: they ignore practically irrelevant phenomena such as turbulence in a water pipe system. They furthermore introduce knowledge that is not part of or derivable from any natural science, as emerged from the discussion on technological specifications.

The selection of relevant knowledge is closely linked to the aim of acquiring technological knowledge. Technological knowledge is not, like scientific knowledge, directed at finding the truth or increasing understanding. It is more directed at usefulness. In Vincenti's words: '…the objective of the mission of the engineer [is] to design and produce useful artefacts. By contrast, the scientist seeks to gain a knowledge of the workings of nature' [Vincenti, W., 1990, p. 131].

Underlying this is the idea that engineers are never concerned with knowledge-for-knowledge's-sake. Many engineers and philosophers who contemplate the nature of the work of engineers subscribe to this idea, which seems more defensible than the idea that for engineers the developing of knowledge is immaterial (see Section 4.1).

Three complications arise for the idea that there is a fundamental distinction between "usefulness-first" technological knowledge and "truth-first" scientific knowledge.

The first is that it is customary at many technological universities to distinguish engineering *research* from engineering *design*. This distinction demonstrates that there are engineering disciplines in which the central concern is not to design artefacts, but rather to generate knowledge about artefacts through research. Perhaps all such knowledge ultimately finds some practical application, but there need not be a direct link with a particular useful artefact. To return to an earlier example, control-volume analysis is a general method that cannot be directly judged on the basis of, for example, its usefulness for designing propellers. Rather, one of the reasons why control-volume analysis is deemed so useful is that it is based on physical knowledge that we take to be true; one might say that it is, in effect, a tool for making true knowledge useful. This holds for many more examples of useful knowledge, which are based on – but not identical with – true claims about the world. Knowledge can, of course, be useful without being true, as all sorts of unrealistic but convenient models demonstrate. But it is undoubtedly true that engineers strive after knowledge that is *both* useful and true, even though they might find usefulness more significant if they cannot have both usefulness and truth.

A second complication is that usefulness also plays an important role in scientific knowledge. According to so-called 'instrumentalists' in the philosophy of science, scientific theories are not accepted because they might be true but rather because they are useful for predicting and possibly describing and explaining the results of measurements. Even those who do not want to go this far cannot deny that nowadays possible applications play a major part in assessing scientific knowledge, even of the most fundamental kind. When writing proposals for research grants, scientists are currently required to indicate the scientific and societal significance of the proposed research project. Allegedly, only some 25 years ago, mathematicians could fill in 'not applicable,' but in this day and age, such disregard for possible applications would result in rejection of the proposal.

If you are still convinced that there is a fundamental divide between scientific and technological knowledge, then here is a third reason to doubt that idea. Science and technology have, over the last two centuries, become ever more closely intertwined. One of those intimate links is seen in the area of experimentation. In the year 1820, Oersted was still able to produce pioneering results with a compass needle and a battery. Perhaps low-tech groundbreaking research is still possible, but the vast majority of present-day experiments are multi-million feats of engineering with many of the complications that also plague, for instance, large-scale infrastructural projects (think, for instance, of the *Large Hadron Collider*, the CERN particle accelerator in Geneva). Another place where science and technology are hard to tell apart is in the storing, processing and distribution of information, and in the communication and cooperation between academics. Here, an ever-increasing use is made of ICT, which has sometimes been especially developed or adapted for these particular purposes. Beyond this, an increasing number of technologies are being used to resolve scientific problems or at least facilitate the search for solutions. Some examples are simulation techniques that are used to test hypotheses and models, and approximation methods that can only provide sufficiently accurate results if they are processed on powerful computers. Institutionally, too, science and technology are, to an increasing degree, becoming intertwined. A case in point is the practice of applying for patents for scientific outcomes.

In philosophy, history and sociology of technology, many chose not to write of 'science' and 'technology' anymore but instead of 'technoscience'.[23] The term refers to any kind of scientific research that is conducted in a technological context, and that cannot be understood in isolation of that context; and to any technological achievement that cannot be separated from scientific research. Work on technoscience has, for instance, charted its history and recent development into large-scale

---

[23] Influential descriptions of 'technoscience' are to be found in the work of the philosopher of technology Ihde, D. [1979, 1991] and in the writings of the sociologist Latour, B. [1987].

technoscience projects such as particle accelerators; others have examined how advances in science after World War II were made possible by financing from military institutions and how they, in turn, contributed to advances in military technology. There is no fundamental distinction between scientific and technological knowledge in this research; in fact, trying to make such a distinction is regarded as counter-productive.

All the same, the idea that truth and usefulness play different roles in technological and scientific knowledge is very credible. Still, the above-mentioned problems show how hard it is to formulate the idea as a one-liner. Technological knowledge might be less concerned with truth, which is why engineers are prepared to settle for knowledge that 'works' whereas scientists are supposed to pursue knowledge that is 'true'. In addition to this, technological knowledge has a more direct link with practical concerns than scientific knowledge. If scientific knowledge is connected with a certain artefact or has some direct practical implementation, one might describe it as technological knowledge 'in disguise'. That it was developed by a scientist does not make it scientific knowledge, just as physicists might attempt to solve problems in economics. Control-volume analysis shows that the connection with a specific artefact or a specific practical goal can be loose. The best conclusion might be that there is a sliding scale with nuance differences, depending on the domain and goal.

An inventory of the theories and models that engineers use provides evidence for this sliding-scale view. A certain portion of those theories, such as classical mechanics and thermodynamics, is purely scientific. Other theories, such as control-volume analysis, are derived from scientific knowledge but simultaneously reveal interests in practical applications that are the province of engineers. These interests are, for instance, notable in certain details that tend to be ignored. Still further away from the scientific end of the spectrum, one can find phenomenological theories. Such theories are based on presumptions about a given system that are indispensable for calculation purposes but which are clearly incorrect. Engineers (*and* scientists) turn to phenomenological theories especially when no 'correct' theories are available. That is, for instance, the case in situations where turbulence plays a prominent role. Finally, some technological knowledge is systematic but can no longer be described as theory. In propeller design, it is impossible to rely on information drawn from theoretical models. Still, designers need to have information about the behaviour of propellers in order to make design choices. Such data is obtained from extensive tests done with real propellers and with scale models in which use is made of parameter variation and dimension analysis in order to 'translate' the test data to real data on the actual performance of propellers. Here, more prominently than with phenomenological theories, practice dictates how knowledge is accumulated. A scientist would probably not wish to stick his or her neck out by making claims about how a propeller might behave under complex circumstances, but an engineer does not have a choice. If a scientific theory were to produce practically applicable data no engineer would, of course, refuse to use such a theory. As it is, as long as no such theory is available, design choices have to be based on other sources. It would be sheer scientific chauvinism not to describe such alternative sources as 'knowledge'.

c. Technological knowledge has a 'know-how' facet.

Philosophers, following Ryle, G. [1949], often distinguish two types of knowledge. One is the kind of knowledge that can be expressed entirely in terms of assertions; the other is not or, more to the point, cannot be expressed in such a way. This differentiation is formulated in at least two different ways. In many languages, the word 'know' is used to express two different kinds of things. First, there is 'know that' as in 'I know that a plane flies because its wings create lift' and 'know how' as in 'I know how to construct an operational plane'. This linguistic distinction is often extended by proposing that 'know how' assertions cannot be entirely traced back to 'know that' assertions. The standard example of this is cycling. Perhaps you are reasonably good at describing, in language, what is required to ride a bicycle, but could you teach someone to do it via a correspondence course? Probably not. Cycling is typically something that is learnt by *doing*. This suggests that 'know-how' or 'practical knowledge' cannot be completely reformulated in the theoretical or propositional knowledge[24] that fills textbooks.

A close cousin of know-how, perhaps even the same thing by a different name, is implicit or tacit knowledge.[25] This knowledge is also gained from personal experience and cannot be verbally conveyed to others. Imagine, for a moment, that you are an experienced pilot who has flown all types of aeroplanes. You can partly make those experiences explicit, also in the form of assertions, but communicating them precisely across to others is difficult and will always remain sketchy and fragmentary. Teaching someone else how to fly will only work by personally supervising him or her – but perhaps even that is not enough, as you can only become an experienced pilot by flying a lot *yourself*. This partly explains why it was so hard to draw up technological specifications for the navigability of a plane: pilots found it hard to verbalise their experience in this field.

The ideas on 'know-how' and implicit knowledge have drawn a great deal of attention, also in the literature on knowledge management. If you wish to transfer the knowledge that is available within an organisation or company from one member to another, for instance, because of staff changes or because information has to be shared with other branches, then you would do well to know how such knowledge transfer works and how limited written instructions might be as a means of communicating all the relevant knowledge. The role of tacit or implicit knowledge could clarify these limitations.

Psychologists, management scientists, economists and philosophers therefore enter into extensive debates on the part played by 'know-how' and tacit knowledge in all kinds of processes. There have also been high expectations for clarifying the 'tacit dimension' of technological knowledge. Some authors, such as Schön, have emphasised that designers perpetually make use of highly personalised and difficult to transmit forms of knowledge and experience (see also Section 2.3). Besides, just as the example of cycling makes explicit, the use (and design) of artefacts contains an element of 'know-how,' which people rightly accept cannot be completely put into words.

---

[24]A proposition amounts to the content of an assertion, for instance that 'the earth rotates around the sun'. 'Know-how' cannot be termed propositional knowledge if it cannot be completely expressed in terms of assertions.

[25]This idea was elaborated by the chemical analyst Polanyi, M. [1966], who also stressed the importance of operational principles.

There are at least two forms of technological knowledge that have a large implicit component. The first is the type of knowledge that literally cannot be expressed in words because it is captured in images. Means of visualisation, from sketches to CAD simulations, are vital to engineers and cannot be replaced by text. It is unthinkable that a new aircraft might be designed without the aid of visualisations in every step of the process. Another type of knowledge that is highly implicit is common sense or the capacity to make practical judgements. Since engineers solve practical problems they are involved in complex, changing circumstances, they work under time pressure and they face uncertainties. Artefacts cannot be tested under all conceivable circumstances; and sometimes there is no theory to adequately describe, let alone predict, their behaviour. Implicit knowledge, gained from practice in the field, can help engineers – and is often essential – when it comes to deciding what risks to take or uncertainties to accept instead of carrying out further tests or developing more accurate models.

This is not only true for engineering: 'know-how' and practical judgement play a major role in virtually all terrains, also in the natural sciences. Just like engineers, scientists gain all kinds of practical experience during their education and when conducting research. That experience is put to good use when they set up experiments (which can be seen as a technical aspect of scientific research) but also when it comes to solving mathematical problems, and constructing models and theories. Research conducted into scientific problem solving strategies has, for example, proved that scientists very often visualise their problems and draw on analogies with more familiar problems.

d. Technological rules and use plans

The final feature builds on the previous two and adds a further crucial element. Technological knowledge must be primarily useful (Section 4.3 b), and it is in part practical 'know-how' and knowledge gained from experience (Section 4.3 c). So perhaps it can be partially described in terms of the way in which it contributes to successful actions. Just think, for example, about who would count as a technological expert, as someone who possesses extensive technological knowledge. Those who possess scientific knowledge are predominantly highly educated scientists who pass on bits and pieces of their knowledge (often without the necessary substantiation) to scientifically uneducated 'laypersons'. With technological knowledge, the roles are different. Designers and makers of artefacts need to possess this knowledge, but it is also indispensable for users. Moreover, the role of designer, maker and user are defined in terms of *actions* and not primarily in terms of whether or not they possess knowledge or a particular method of acquiring knowledge – unlike the role of scientific expert. This suggests that actions might also constitute the *content* of technological knowledge: such knowledge is all about the means (actions) required to achieve certain ends.

One way of fleshing out that idea is by asserting that the engineering sciences produce *technological rules*, that is to say, instructions to carry out a certain series of actions in a certain order with an eye to achieving a certain goal. Aviation specialists produce, for example, instructions on how to fly and maintain a plane by carrying out certain actions.

People perpetually give each other instructions. For instance, you can explain to someone else how they can get home by means of the public transport system or how to get rid of a mole in the garden. What might be special about the rules that engineers produce? Perhaps it would be useful to bring back the applied-science-view and defend that technological rules, unlike common-or-garden tips or the contents of the railway timetable book, are indeed based on applied scientific knowledge.

The idea is not very viable, though, partly because many technological rules lack such a scientific basis. The designers of the first steam engines could not depend on scientifically grounded rules; in designs where turbulence has to be accounted for, that is still impossible. Furthermore, engineers use all kinds of *rules-of-thumb* whereby interim results are construed and checked. Sometimes such rules are later substantiated or replaced by substantiated rules but definitely not always.

Another way to distinguish technical rules from more everyday instructions is offered by the use-plan analysis that we presented in previous chapters. The characterisation of technological rules that was given above is very similar to our earlier characterisation of use plans: goal-directed series of actions, which include the manipulation of a technical artefact. And as was previously outlined, designers do not just produce artefacts, they also develop use plans which they communicate to potential users. Every artefact is embedded in such a plan. Knowledge of use plans is of crucial importance to users: without a plan you might know what an artefact is for but not how you must operate it to achieve that goal. A plan therefore has an instructive (prescriptive) character: it contains suggestions for actions that are, to put it mildly, strongly recommended to users.

By communicating use plans, designers therefore transfer practical indispensable knowledge to others. This is no factual knowledge but knowledge about rules, required or recommended actions. Designers do not need to ground such knowledge in scientific theories. Engineers can test prototypes and adapt them on the basis of parameter variations or trial-and-error; in that way, they might design a 'tried-and-true' combination of artefact and use plan that is spectacularly successful without being able to explain this success with scientific knowledge.

Use plans show that, to a certain extent, technological knowledge must be *procedural* or *prescriptive*. As a designer, you have to do more than just describe an existing situation or a new artefact. You have to know how that artefact should work and what actions people should undertake to realise their goals with the help of that artefact. Instructions on how to land a passenger aircraft do not constitute descriptions of actual landings, though most landings will hopefully be in line with the instructions. Instructions dictate how something *should* happen and therefore constitute a standard for assessing the behaviour of artefacts and people. An engine that misfires does not operate optimally, just like a pilot who does not maintain a safe distance from other aircraft. Even rules-of-thumb are often described in prescriptive terms: if a design does not comply with rules-of-thumb, then it is usually concluded that a calculation mistake has been made or that a wrong decision has been taken in the design process.

By contrast, natural scientists produce no instructions or rules, and the standard examples of scientific knowledge are not prescriptive. Newton's laws of physics do not dictate how bodies should move, they describe what they actually do. Technological rules and use plans could thus well be a unique component of technological knowledge, which helps to distinguish it from scientific knowledge.

4.4   CONCLUSION

We have shown that engineers develop knowledge and that this technological knowledge has a number of specific features and forms. Technological knowledge is partly artefact-oriented, directed at usefulness, tacit and prescriptive. It contains elements that resemble scientific knowledge yet differ in subtle ways, such as control-volume analysis, and elements that cannot be found anywhere else in the natural sciences, such as knowledge about operational principles and rules for use and design.

This demonstrates that the traditional 'applied-science' image is wrong: engineers can not and do not simply apply scientific knowledge; they are knowledge producers as well as consumers. Yet it is difficult to make a fundamental distinction between scientific knowledge and technical knowledge. The similarities seem to be too great for that. Moreover, science and engineering are nowadays so interwoven that it would not be realistic or even possible to keep apart the knowledge produced in these activities. Still, it might be fruitful to analyse in more depth and detail the features of technological knowledge, such as its 'know-how' nature, and the part played by rules and use plans: if all our knowledge had all those features, it would only become more necessary to improve our understanding of these features and thus of our own knowledge, technological or scientific.

4.5   A FEW MORE ISSUES

We have examined a number of the features of technological knowledge, such as its connection with actions and with usefulness, yet we do not fundamentally differentiate it from scientific knowledge. Establish whether that latter idea is right by looking for concrete examples of scientific knowledge that share some of the characteristics of technological knowledge. What knowledge about artefacts can, for instance, be said to be scientific? What kind of scientific knowledge has a large tacit component but at the same time a clear link with rules and actions? In what kind of scientific knowledge does usefulness play a major part? Determine whether these examples of knowledge could be termed 'scientific' or would it be more apt to call them 'technological' (even though it is perhaps part of physics).

A more difficult issue is whether the particular characteristics of technological knowledge can be emphasised without implying a fundamental distinction with scientific knowledge. One way this might be done is by placing the two forms of knowledge on a sliding scale going from 'very scientific' to 'purely technological' without drawing a sharp dividing line somewhere on the scale. Try to see if that actually works by placing the following types of knowledge on such a sliding scale: fluid dynamics, the knowledge of Vincenti's pilots, the knowledge of a good model plane builder, the knowledge of a team of aeroplane constructors at Boeing and knowledge about fibre materials such as Glare. Do these kinds of knowledge feature on points of the scale or in intervals? Is one scale sufficient?

Furthermore, think about why it might matter to make a difference between scientific and technological knowledge or science and technology. Why might people want to do this? Do they really need a fundamental and sharp dividing line or would a sliding scale work just as well?

CHAPTER 5

Sociotechnical Systems

In Chapter 1, we argued that technical artefacts such as aeroplanes, electric drills, computers and ballpoint pens differ from both physical objects and social objects in that they embrace something of both. Technical artefacts are tangible objects with physical properties, but they are also objects with a function, which they have in virtue of their embeddedness in use plans aimed at the achievement of human purposes. In Chapter 2, we examined the way in which technical artefacts come into being, by way of a design process, and in Chapter 4, we considered the knowledge that this requires. In this chapter, we show that to view technology as merely a 'collection' of technical artefacts would be an immense oversimplification. In so doing, we would completely fail to acknowledge the layeredness that is such an important feature of modern technology: the technical artefacts discussed so far are building blocks in wholes of a far greater complexity. Although one cannot build 'loose' technical artefacts that are as big as the earth itself – where would one be able to assemble them? – the wholes or *systems* central to the present chapter do, in fact, span the entire globe. We shall see that as a result of the character of these sorts of systems, the principles traditionally adhered to by engineers when designing technical artefacts, and the kind of knowledge upon which they rely in the process, cannot remain unaltered when it comes to the matter of the designing and implementing such complex things. As has already been noted in Chapter 3, this also has consequences for the ethical dimension of technology.

5.1   HYBRID SYSTEMS

Imagine that you are walking around a major airport because you are about to travel to a distant destination. You will make use of numerous artefacts. Some of these, like the baggage trolleys and the benches in the cafés and waiting areas, will be simple artefacts, probably designed by just one or a few persons. Others will be incredibly complicated, designed by whole teams of engineers who will probably have worked on them for years, for example, the computers used for the checking-in system where the details of all the flights are stored and, of course, like the showpiece of engineering ingenuity, the actual aeroplane itself in which you will ultimately fly. However, alongside all these separate things you will also be making use of something much more encompassing, just as much created and maintained by human hands as the prototypical artefacts just mentioned, but at the same time, it is much more impalpable and harder to fathom. That 'thing' is the *world civil aviation system*. All the artefacts just mentioned constitute a part of it, but there is incredibly much more to it than that. Some of the components are concrete, like the buildings in which the passengers are subjected to a whole range of routine procedures before entering the plane or after stepping out of it, and the runways used for take-off and landing. Additionally all kinds of equally tangible people are involved: the cabin crew, of course, but also the personnel who work at the check-in desks, operate the X-ray equipment, check the passports and load the baggage onto and off the plane. But then, there are also, in very divergent ways, numerous other components of a more abstract nature: the air corridors within which the aeroplanes fly, the airline companies responsible for the flights, the regulations that pilots and airline companies have to observe, the organisations that draw up and enforce these regulations, the treaties agreed to between countries that make it possible for planes to fly from one airspace zone to another, the companies that insure the system for the different ways in which it can fail, and so forth, and so forth. Each of these things contributes to the functioning of the world civil aviation system. Without this system, it would be impossible for you to travel by plane to your chosen destination and afterwards to fly back home again. Without this system, you would undoubtedly not find at your final destination the places justifying the trip in the first place, like a holiday hotel, a conference resort or a business centre. Without each of the listed components, the system would not be able to function in the way it does at present, certainly not on the scale and with the level of efficiency that most of us completely take for granted.

The world civil aviation system is an example of a *sociotechnical system*. The fact that we refer here to a system will not come as a surprise to most. A system is understood to be an entity that can be separated into parts, which are all simultaneously linked to each other in a specific way. An aeroplane is itself a system. All of its components, however, are 'hard' things, the behaviour of which is governed by various natural laws. A thorough knowledge of these laws is required in order to comprehend how an aeroplane operates from the way in which its components work together and to design and connect the components in such a way that the resulting plane does precisely what is expected of it. Admittedly, in present-day aeroplanes, there are all kinds of control systems that are computer-steered, running on software systems that were designed in isolation of the computer and, in that respect, definitely not tangible. However, once downloaded, this software ensures that the onboard computers are set up in a specific physical state, which is essentially no different from the way in which a thermostat is adjusted to a specific setting for it to maintain a certain temperature.

The presence of software thus changes nothing whatsoever about the character of the aeroplane as a physical entity. What makes the aviation system a special sort of system is the fact that it includes all kinds of components – the organisations and institutions, the conditions and rules – that are not tangible things. For all of these components, no thorough knowledge of the natural sciences will do to understand how they work and to fit them into the system in an effective way. The relevant employees mentioned, in their capacities as human beings made of flesh and blood, are tangible, but that is hardly relevant to the position that they occupy within the system. One does not need to possess knowledge of biology to fit the staff working at the check-in desks into the system as a whole. The most basic everyday knowledge is sufficient to do justice to their character as physical-biological organisms that cannot walk on air or pass through walls, so that they need accessories like doors, seats and floors to be able to carry out their work. Which is not to deny that occasionally, specialised knowledge about the biological side of humans is relevant, for instance, when designing cockpits where a pilot's attention has to be divided between a wide range of instruments, or – to briefly step outside the framework of civil aviation – when designing fighter planes where the pilots are exposed to extreme acceleration speeds.

In the case of people employed within the civil aviation system, matters such as their height, weight, gender or stamina are not the issue at stake; what counts is the fact that they are *persons*, capable of understanding and carrying out instructions and also of understanding the purposes served by these instructions. The regulations and the organisations and institutions that contribute to the aviation system presuppose the status of human beings as persons. Rules and regulations are drawn up by people and can be observed or flouted by people. Organisations and institutions are created and maintained by people. In order to assess the way in which people function within the civil aviation system, we largely rely on a general picture of the way in which people go about doing the things they do in everyday life, supplemented with knowledge obtained from the fields of sociology and psychology. None of the natural sciences, however, has anything useful to say on the matter. We cannot find either rules or organisations anywhere in the mineral, plant or animal world.

This, thus, brings us to the heart of the matter of what makes the world civil aviation system such a special sort of system: it is a *hybrid* system. It consists of components which, as far as their scientific description goes, belong in very many different 'worlds'. This is what makes them essentially different from even the most complex of technical systems, like for instance civil aircraft. Even though the engineers who were involved in the designing and the manufacturing of the Airbus A380 had very different backgrounds – mechanical engineering, materials science, aerodynamics, electronic engineering and computer engineering – all these disciplines share a form of describing the world rooted in natural science. However, the aviation system into which such an Airbus A380 operates involves numerous other things – people, institutions, rules – about which the natural-scientific way of describing the world has little to say and for which a social-scientific way of formulating matters is therefore required.[26] Hybrid systems, in which certain components, are described and researched using the natural sciences and other components, are described by drawing on the social sciences are called *sociotechnical systems*.

This hybrid character of sociotechnical systems needs to be distinguished from the dual nature of the technical artefacts that was central to the discussion in Chapter 1. This dual nature even applies to the most simple of technical artefacts, such as a screwdriver or a nutcracker: they are all objects that, apart from their physical properties, also have their particular function and their embeddedness in a context of human use plans as non-physical features. Sociotechnical systems, being technical artefacts with an extremely high degree of complexity, are, in principle, also subjected to this dual nature. One can view a sociotechnical system as a particular 'thing' having certain causal properties, and one can examine the function that it has in a context of human actions. By contrast, the hybrid nature of a sociotechnical system has to do with the *composition* of the 'thing' it is, as a result of

---

[26]Unfortunately the realm of human and social sciences is much less well-organised than the realm of the natural sciences. We use here the term social sciences as the most general term for a set of disciplines containing the humanities as well as the social, economic and cultural sciences and also parts of psychology such as cognitive, social and organisational psychology. We shall not deal here with the various interrelations between these specified disciplines.

which it can no longer be unambiguously seen as a single tangible thing to be picked up, in a manner of speaking, and held up to the light for further inspection. This hybrid character makes the examination of its causal aspects a much more problematic matter than in the case of traditional technical artefacts, with all the consequences that this entails for the designers of such systems. It also gives the dual nature of technical artefacts a much more complex character, however. On the one hand, the social aspect, the context of human action, is manifested in every system aspect of the total system within which people are involved. On the other hand, it is for sociotechnical systems, in particular when they reach country or continent scale, hard to identify the function of the system as a whole, as it is simultaneously embedded, at any one moment, in the context of action of numerous different individuals.

Even though it is eminently this hybrid character – the presence of components requiring a physical description and components requiring a social description – that characterises sociotechnical systems, the designing, implementing and maintaining of these systems remains predominantly in the hands of engineers, who have been educated in pronouncedly natural-scientific ways. That is why these systems constitute a major challenge for the engineering sciences. All kinds of traditional notions about what constitutes the designing of a technical artefact, how the design process should be structured, what kind of knowledge is required and how one should assess the functioning of a designed artefact, become very problematic whenever they are literally transplanted to the context of designing sociotechnical systems. The reasons for this are presented in the following section.

5.2   SYSTEM ROLES FOR PEOPLE: USER AND OPERATOR

The special character of sociotechnical systems is not grounded simply in the interaction between man and machine. Virtually all technical artefacts have an on/off knob for the purpose of starting or stopping their functioning, and on top of that all kinds of knobs, switches and handles allowing the user to adapt their functionality. Technical artefacts are, after all, manufactured to be used and this presupposes that there is a *user* who makes use of the artefact by manipulating it. That does not mean that using an artefact necessarily requires continuous manipulation. Sometimes the manipulative aspect only extends to the installing of the artefact, inserting it into some network of causal connections, like when a memory chip is installed, a smoke detector is connected or a communication satellite is launched. In the artefact's use plan discussed in Chapters 1, 2 and 4, the kind of manipulation suitable for the artefact in question is specified. But without some form of physical interaction, we can have no use. What makes sociotechnical systems special is, first of all, that they have many users at any one moment and, secondly, that they involve people in two different ways, namely, not only in the role of user of the system but also in the role of operator. The word 'role' is apt because one and the same person can simultaneously be a user of and an operator in a system. If a pilot plans to spend some time at his or her destination for a holiday, after having flown the plane there, he or she is, on the one hand, performing the role of operator by flying the plane and, on the other hand, he or she is performing the role of user by using the aviation system to get to his or her holiday destination.

The role of operator may well seem to appear out of the blue. Does the pilot, in his or her role as flyer of the aeroplane, really do something different from what the owner of a coffee-maker does when turning the machine on to make a cup of coffee? There is, indeed, an important difference: the user of the coffee-maker operates the machine to realise a goal of his or her own that would be much harder or even impossible to realise without the device. The pilot of an aeroplane that is on its way to Singapore does, we presume, have instructions to arrive there within a reasonable period of time, but he or she is not doing so in order to realise his or her own goal of arriving in Singapore within a certain amount of time. It is the passengers who have this as a goal and by taking that plane they realise that goal. To do so they not only use the aeroplane but also the entire aviation system, which they 'operate' by purchasing a ticket. The pilot of the plane is a *component* of that system, a component that is necessary for the whole system to be able to fulfil its function, just as much as the plane that the pilot flies is. After all only few of us have at their disposal their own private air transport to take them all the way to Singapore, and the operation of civil aviation aircraft is too complex to be left to the travellers themselves. It is also (as yet) too complicated and too expensive to manufacture an aeroplane that is capable of operating completely automatically, taking off without any human intervention after the last passenger has boarded and closed the door behind him- or herself, and then landing all the passengers safely in Singapore some ten hours later.

Pilots may be the most obvious indispensable human link in the civil aviation system, but they are certainly not the only ones. Just as indispensable are the air traffic controllers who sit in the control rooms at airports and supervise the taking off, landing, cruising and taxiing movements. The great speed and limited manoeuvrability of civil aircraft, combined with the limited vision of pilots would make all air traffic involving more than one plane in the air at any given time virtually impossible if there were no air traffic controllers to maintain radio contact with the pilots and to monitor, by means of their radar equipment, the position of all the aeroplanes. In much the same way, all the staff members mentioned above – the staff working at the check-in desks, those who operate the X-ray equipment, the passport controllers and the ones who load the baggage in and out of the holds – are components of the aviation system, each contributing a specific function to the operation of the entire system.

All sociotechnical systems have operators who fulfil such roles, because it is too difficult or even impossible to build a system consisting merely of interconnected technical artefacts – that is, machines or 'hardware' devices – and guarantee its adequate functioning. Every large chemical plant or power station has control rooms that are permanently occupied by one or more operators or controllers. Even in road transport systems – which, even though forming another sort of transport system, are very different from aviation systems, as we shall see – operators are coming to play an ever bigger part, with the increasing volume of traffic on roads, by monitoring the flow of traffic and attempting to control it, through the imposition of speed restrictions, the opening or closing of traffic lanes and the provision of specific information.

Human operators started to receive serious design interest with the rise of *systems engineering* during the first two decades after the Second World War. At the instigation of many military organisations engineers were increasingly becoming involved in the development of complex systems consisting of numerous components of a divergent nature in which knowledge drawn from very different disciplines was processed. The main challenge facing those responsible for the overall design process was to coordinate the behaviour of the separate components. Since the development of the computer was very much in its early phases, the only 'mechanisms' available that were capable of realising such complex coordinating activities were human beings. Thanks, though, to the rapid pace at which the computer has developed since, human beings are becoming increasingly redundant. The control rooms of factories and plants have thus become gradually emptier and within the foreseeable future the fully automated flying of aircraft will be a reality.[27]

This should not, however, be taken to mean that large-scale systems involving human operators will be merely a temporary phenomenon. The complexity of sociotechnical systems, particularly large infrastructural systems, consists just as much in the fact that, unlike a typical technical artefact such as a coffee-maker, they have very many different users. The functioning of the system as a whole, as it appears to each of its users, not only requires coordination between the technical or *hardware* aspects of the system and the behaviour of the users – like the driving of the vehicles by their chauffeurs – but also, and especially, the mutual coordination of the behaviour of the many users. To achieve this coordination, it will not to do to implement one or another causal mechanism that is attuned to the physical characteristics of the users. Successful coordination comes about through agreements, rules, laws, habits, in short, precisely the sort of things that are studied in the social sciences and not in the natural sciences.

5.3   RULES AND COORDINATION MECHANISMS

It is in the notions underlying agreements, rules, laws, and so on, that one becomes most sharply aware of how it is that sociotechnical systems differ from traditional technological systems. If you want to 'direct' people, it is common to do this through *rules* or *instructions* and not through causal stimuli and signals.[28] A *rule* is a directive or norm that has the underlying intention of bringing about a behavioural pattern, irrespective of whether that pattern actually occurs. A rule can be observed or ignored, just adhered to from time to time or abided by depending on the circumstances. If a rule is not followed in a particular case, this indicates that it is apparently not in everyone's best interests to behave in the manner that the rule dictates.[29] Such deviations will not necessarily undermine the intended behavioural pattern, but they are likely to do so, especially if many people are tempted to breach a rule because acting according to the rule requires an effort or is costly in one form or another. What is especially characteristic of a rule is, therefore, the existence of *sanctions* relating to the breaching of the rule. As a result, rules also presuppose the existence of a social group in

---

[27] See, for instance, Weyer, J. [2006].

[28] To be sure, direction by way of causal stimuli and signals is occasionally applied, and successfully, but it requires precisely engineered, stable circumstances to be effective.

[29] We distinguish, therefore, between a rule and a *convention*, where a convention is an actually occurring behavioural pattern, which has come about through mutual coordination of behaviour among different people and which continues to exist, even without being prescribed, for the simple reason that it is in everyone's interests to abide by that convention.

which the rules are considered to apply and through which sanctions can be enforced. After all, an individual cannot be expected to enforce sanctions upon himself if a rule is deviated from.

In order to allow a sociotechnical system to function properly, therefore, rules have to be thought up or drawn up and imposed as a coordination mechanism. Consequently, these systems take on a kind of complexity that is lacking with technical systems. The role of an operator within a sociotechnical system is defined by a set of rules or *instructions* that determine in what situation the operator must take what particular measures. These instructions constitute, as it were, a – more or less forcefully prescribed – use plan for those components of the system that the operator has under his or her control. In addition to this, there are use plans for the users of a sociotechnical system that tell them how they may and must use the system and how they have to behave when using the system if it is to retain its functionality for others. Just how the system as a whole will function depends crucially on the way in which these rules are formulated. On the one hand, the actions that an operator or user carries out, in accordance with the rules, must indeed give the results corresponding to the functionality of the system, given the actions that these same rules impose upon other users. On the other hand, the rules must be of such a nature that it may reasonably be expected that operators and users will actually carry out the actions that have been stipulated in them. In other words, once laid down rules must also be followed. This latter aspect is irrelevant in the case of purely hardware systems.

The physical behaviours of the *hardware* components of a system follow law-like regularities that can be researched and established by means of experiments and tests. As long as there is uncertainty concerning the physical behaviour of a component no engineer will be prepared to include it in the design. To be sure, it is inevitable a matter of judgment whether or not there is sufficient knowledge available about the behaviour of a particular component or material, and unexpected things leading to the failure of the system can never be completely excluded. Still, when designing technical systems, an engineer can rely on the existence of natural laws at various levels. Engineering design is completely based upon verifiable reliability. This extends to technical systems whose functionality is supported by software: once loaded with a certain program, a computer system will go through a sequence of states in an entirely predictable way, and provided that the software is free of errors, these states will be precisely the states that a component must have, according to the design, to make its contribution to the realisation of the functionality of the entire system.

At first sight, the situation that applies to sociotechnical systems does not appear to be significantly different. You could 'calculate' what behavioural dispositions must be given to the people within the system if you want the system, including its human components, to function as intended. These behavioural dispositions could subsequently be inculcated in the relevant people, either through training or purely by instruction. This could be seen as a process of 'loading' a certain 'program' into a person. If we could assume that an individual will behave exactly in conformity with a set of behavioural rules, once 'loaded', then the design problem will be restricted to the 'designing' of these dispositions or instructions, which is not very different from the process of designing software for the components of a purely technical system. The problem, however, is that this situation, in which people will unfailingly execute specific instructions and completely and meticulously follow instructions, seldom or never occurs, and can hardly be expected to occur, in view of the constitution of the people required to execute or follow the directives. Every individual is, as it were, a computer on which a great many programs are running all the time, programs about which the designers of the sociotechnical system in question have only a very general and limited idea. Rather than loading operators with a single program that secures the precise execution of the operator's role as the one and only thing that the operator cares about, designers can do no more than add to the vast amount of already installed software a mere handful of subroutines.

5.4   SYSTEM DESIGNS AND SYSTEM BOUNDARIES

With this understanding, we come to the problems that confront those who are responsible for designing, implementing and maintaining sociotechnical systems, and we come to the challenges they pose for the way in which engineers are traditionally taught to tackle the designing of artefacts and systems. These problems will be dealt with from two angles. First, there is the problem of how to draw the system boundaries and, accordingly, to establish the extent of the design task. In the second place, there is the problem of the predictability of the system's behaviour and the extent to which this can be controlled.

The most important thing that can be said about the boundaries of a system is that these are not given beforehand but have to be decided upon on the basis of various considerations. Or, to put it more precisely, it is only once a boundary has been drawn that it is clear exactly which system is the subject of research and design. A system is, after all, a collection of interrelated components, each of which can be broken down into smaller components, which makes these components, in their turn, also systems. Furthermore, every system that we define by drawing a boundary is connected beyond that boundary with other things so that we can view the system under investigation and these other things as components of a still wider system. In the case of natural systems, this series of embeddings spans the entire spectrum from the most elementary particles at one end to the entire universe at the other end. For hybrid systems of the type discussed here, however, the spectrum is obviously much more confined. The smallest component of a hybrid system is one single person or a single technical component while the largest system that could be described as an entire sociotechnical system must remain, for the time being, confined to the earth.[30]

As far as the designing of systems is concerned, the important question is not what boundaries can be drawn in an existing reality, but what place the new system is to occupy in the existing reality? As long as this is unclear, the design task with respect to the system is still undetermined. Imagine that, as an engineer, you are requested to design a new type of aeroplane engine. In order to be able to do that, you not only need to know of which technical system this engine is to be a component – for what type of plane it is required – but also of which (socio)technical system the plane will be a

---

[30]The position of a sociotechnical system in terms of time and space is not a foregone conclusion. There are, for instance, all kinds of spaceships dotted around the solar system, spaceships that are linked to control centres on earth and which, in most cases, can also to an extent be controlled by those centres. Does this mean that our largest sociotechnical system extends to our entire solar system? It is an intriguing question, but we shall not endeavour to address it.

component. If the aircraft is merely destined to fly in the private airspace above the land of a large estate owner (let us presume that something like that is possible), then the functional requirements can be limited to the exclusively technical requirements: power, thrust, weight, and so forth. If, on the other hand, the aeroplane in question is destined to fly as a component of the existing world aviation system, then an accordance with all kinds of legally stipulated norms and standards will also form part of the functional requirements for the engine. An aeroplane that does not meet these requirements cannot be integrated into the system, just as little as a bolt will fit into a nut with an incompatible screw thread. These norms and standards are preconditions to which you, as a designer, must conform in much the same way that designers, in general, have to accept the laws of nature, which they cannot change.

Matters become slightly more complicated when the issue is the design of an entire aeroplane. If what is intended here is a private plane for the same large estate owner, which will exclusively be flown in his private airspace, then again the designing is merely a technical artefact. Naturally you will provide the owner with a manual with instructions for use, or you might even offer him a training course, but, in essence, it will be no different from the instructions for use you would provide for a coffee-maker. After delivery, the owner is then free to vary matters as he wishes, to develop alternative operating methods or to discover by himself how the aircraft can be used for all kinds of stunts. In the case of an aeroplane that is destined to become a component in the world civil aviation system, matters are again different: it now even becomes necessary to decide whether the design task will be confined to just the technical artefact 'aeroplane' or whether it should be extended to the sociotechnical system 'aeroplane plus flight crew'. In the first approach, the responsibility for the drawing up of exhaustive and adequate operational instructions and for the training of the pilots lies with a different 'designer' and, as aeroplane constructor, you are only concerned with external (safety) norms concerning the layout of the cockpit, the instruments and the operation panels. In the second approach the drawing up ('designing') of the rules that define the pilot's role and the coordination between the way the aeroplane itself is designed and the way in which it is operated also form part of your design task. Evidently, the second design approach could produce improvements with respect to safety, which will not be recognised, or not so easily, in the first approach. Which of the two approaches is adopted is in no way predetermined; it remains a question of choice. It is a choice that does not generally lie with the designer but rather with the owners or managers of the (smallest) system within which the aeroplane under design will function as a component.

These examples hardly touch on the complexity of the civil aviation system, which contains a large number of subsystems, each with its own operators. For all those subsystems, separate sets of rules have to be designed. These rules then have to be attuned to each other – especially those that define the role of the pilot and the rules that define the role of air traffic controller – but what must be perpetually borne in mind is the fact that the civil aviation system is embedded in at least one larger system – that of the network of sovereign states. The rules that bind people in their various roles within the aviation system must therefore also accord with the rules that national and international legislation imposes upon individuals. It is a kind of complexity that notably comes to the fore at moments when something goes wrong, which is why we shall now examine more closely a tragic aviation accident that occurred on July 1st, 2002, in the airspace above southern Germany. At an altitude of ten kilometres, a Tupolev 154 from Bashkirian Airlines crashed with a Boeing 757 from the freight carrier DHL. Such an accident always has a whole chain of causes and is, in that respect, the outcome of an unfortunate sequence of events, but from the design perspective, this particular accident demonstrates that also, at the highest levels of complexity, we have to bear in mind the system character of the technology we jointly create.

5.5   THE MID-AIR COLLISION ABOVE ÜBERLINGEN[31]

In 2002, in the wake of previous mid-air collision incidents, aircraft making use of European airspace were all equipped with a TCAS, *traffic collision avoidance system*. In the nose of the cockpit, there is an instrument that sends out a signal but can, at the same time, pick up signals sent out by other aircraft. When the received signal, in combination with the plane's own position and cruising speed, indicates that unless one or both aeroplanes change course, they will collide, the TCAS equipment transmits coordinated instructions to the pilots: one of the crews is given a spoken instruction to descend and the other is instructed to ascend. The TCAS is intended as a last resort in an emergency: it is the task of the relevant air traffic controllers to notice, at a much earlier stage, that two aeroplanes are flying at the same altitude, on courses that will lead to disaster and to rectify the situation by directing one of the two to a different altitude. The TCAS was introduced for situations where air traffic controllers fail to do so. This was indeed the case in the Überlingen incident, but there is more to the story. The air traffic controller on duty in the area where the aeroplanes were flying had indeed failed to notice the impending accident in time and had thus not intervened when he should have. Eventually, however, he did notice the problem and intercepted by instructing the Russian aeroplane to reduce its altitude. But the air traffic controller's instructions came so late that by then, the TCAS on board both aeroplanes had been activated: on the basis of the signals that had been exchanged, the software had generated instructions to the effect that the captain of the American plane should descend while the Russian aeroplane had been instructed to ascend. There was just one second's difference between the Russian captain receiving the message generated by the TCAS to ascend and being instructed by the controller on the ground to descend. Of course, this led to great confusion and debate among the Russian pilots, but there was very little time available for finding out what to do, and after air traffic controller and TCAS had repeated their conflicting instructions, the captain of the Tupolev 154 decided to follow the instructions from the ground and not the message generated by the TCAS. As the Boeing 757 had only been instructed to descend by its own TCAS, this aircraft also started to drop altitude, with the result that shortly afterwards, the two planes collided, causing the death of everyone on board both aeroplanes.

---

[31]For this section use has been made of the official investigation report published by the German Bundesstelle für Flugunfall-untersuchung [2004] and Weyer's book on the subject [Weyer, J., 2006]. Another publication that discusses the role of the TCAS in this particular case, and in general, but which does not view matters from a sociotechnical perspective, is that of Ladkin, P. [2004].

What this accident makes painfully clear is that the TCAS is designed for a closed system involving just two aeroplanes and their crew, but not for a wider system that also includes air traffic controllers. It was designed for situations in which, for one reason or another, air traffic control has dropped out and is no longer involved, but it is implemented nevertheless as a component in a system where air traffic controllers are also present as system components. In the instructions defining the roles of flight crews and air traffic controllers, allowances should have been made for the fact that a crew could receive instructions both from the TCAS and from air traffic control, a possibility that is evident if matters are considered from the perspective of the aviation system as a whole. The existing aviation system regulations, however, provided no answer to the burning question with which the Russian captain had briefly grappled, namely that of which of the two instructions to follow. The whole question ultimately tied up with the embedding of the civil aviation system in the social system of national and international legislation and regulation because during the investigation into the disaster, it emerged that a pilot could refer, in such cases, to no less than five different documents, the exact status of which remained vague and which were not, moreover, in unison with each other. Since this tragic accident, the instructions within the aviation system have been amended by emphatically stipulating that whenever a flight crew receives contradictory instructions from the on-board TCAS and the traffic controllers on the ground, they must ignore the instructions from the ground and follow those issued by the TCAS. But this ruling still appears to ignore the embedding of the world aviation system in the overall social system. In situations where, for instance, three planes are flying in close proximity in the same airspace the TCAS might well fail, either because one of the three planes has no TCAS due to its being, for example, a private or a military plane, or because the software algorithm of the TCAS is not correct, for it has been proved correct only for two-plane situations. Aided by their radar equipment, the air traffic controllers would have the power to correctly direct matters in such situations, but the new ruling obliges captains to abide by the TCAS, even in cases where the captain has good reason to doubt the instructions generated by the TCAS on the basis of his or her own observations. This new ruling therefore, it is asserted, contravenes international legislation, which lays down that the pilot is at all times fully responsible for the safety of the passengers and staff in the aeroplane that he or she flies.

Apparently, then, it is extremely difficult to guarantee coordination between system components up to the highest levels of complexity. What severely aggravates this problem is that such large-scale sociotechnical systems, in fact, have no designers. Rather they evolve through historical processes of spontaneous and directed linkages of subsystems that are integrally designed by engineers or, more to the point, by teams of engineers. If such subsystems become connected, the links between them are adjusted or possibly even replaced by new links, but there is no single organisation which, from a design aspect, contemplates the entire system and investigates whether the functionality of the system is guaranteed by the way in which the components engage. Even a technical system that is developed on a world-wide scale like, for instance, the American GPS system and its intended European counterpart Galileo, is quickly integrated with other existing and newly created systems once it has been implemented, so that yet other systems with new functionalities can emerge for purposes of, in this particular case, telecommunication, navigation and cadastral registration. But also the embedding in the relevant national and international umbrella system is not appraised from a total all-round perspective – simply because there is no authority that represents such a perspective – but rather from the angle of subsystems each with their associated vested interests.

5.6   SYSTEM DESIGN AND CONTROLLABILITY

This brings us to the second problem that sociotechnical systems pose for the traditional engineering approach to design, namely a loss of predictability and control. When developing technical artefacts, the external circumstances within which the system has to fulfil its function are explicitly included in the requirements. Traditionally, it is the task of the designer to produce an artefact that functions as long as the circumstances obtain as specified. Whether the circumstances within which the system is used or implemented, in fact, meet this specification is not so much the designer's problem but rather that of the customer or user, even though, as a designer, one should make certain that the functional requirements taken to define the design task are a correct translation of what the client or commissioner has in mind for the system. With sociotechnical systems, this is completely unattainable. The overarching social system within which every sociotechnical system functions as a component is in a perpetual state of flux. Even if we imagine that a particular large-scale sociotechnical system – say, the world civil aviation system – is developed all in one go, it would still be impossible to precisely specify the institutional context within which that system has to function. This problem that can hardly be resolved by broadening the definition of the design problem and expanding the system borders by involving the institutional context. There is, ultimately, an overarching system, which is the system of sovereign states, but this is a social system, not a sociotechnical system, and at that level, one cannot speak of designing in the sense of engineering designing. The institutional context of national and international legislation and regulation has come into being and functions in a completely different way from technical artefacts. This is even unavoidable because society as a whole cannot be abandoned in favour of a new design; designing is an activity that occurs within the context of society. In that respect, society is like a ship that has to be repaired while sailing on the high seas but kept afloat in the meantime, to use a metaphor introduced by the philosopher Otto Neurath.

The source of the uncertainty with regard to the functioning of a sociotechnical system like the one being considered here is shared by all such systems, though not always to the same degree. The problem was felt acutely during the past two decades with respect to the energy infrastructures, when deregulation turned the institutional context of the existing infrastructure upside down whilst the public expected of the managers of these infrastructures that the functionality of the system be preserved. This situation created major challenges for the operators because it forced them to alter their view of the nature of the system that they are managing and to create new conceptual frameworks and models. There is, however, another source of uncertainty, which hardly comes into the picture in the aviation sector but is very evident in, for instance, the road transport system. In the air transport system, the end-users or passengers are hardly capable of influencing the behaviour of the system. For the behaviour of the aeroplane and for the way in which the pilot and the air traffic controllers go about their work, it does not really make much difference if their aeroplane is loaded with passengers or with freight. It is a system which, in its day-to-day operations, lies close to the engineer's ideal of a completely controllable and predictable system, into the running of which the human operators, through their punctual execution of seamlessly coordinated instructions, smoothly fit.

The same definitely cannot be said of the road transport system. The individual drivers of all the different cars and lorries and the motorbike riders each fulfil a double role: they are users of the system, but they also operate a small part of it. By using the system, a car driver also simultaneously modifies it by introducing a temporary technical component in the form of a car and herself as an operator of that same car. These operators are tied to much less stringent rules than airline pilots and have much greater freedom. The rules that define the driver's role consist, to a large degree, of instructions to react in certain ways to certain signals such as traffic lights, road signs, road markings and the lay of the road. By automatically (e.g., traffic lights) or occasionally (e.g., incidental speed limits and lane control on motorways) altering these signals, the global system controllers attempt to adjust the system in such a way that the constant changes in the configuration of the system are compensated and its functionality maintained. To what extent their efforts are actually successful is a matter of dispute. The fact remains that because of the very nature of the system, the global system controllers must always be uncertain about the degree to which their endeavours will succeed. Sociotechnical systems have unavoidable *emergent properties*, that is to say, properties that admittedly emanate from the properties of the components and from the way the system is structured but which are not predictable, for the simple reason that in order to predict them, one would need to have access to knowledge which, at least in practical terms, is unavailable or, if in principle available, cannot be accessed in the available time.

There are many different schools of thought on what is the best way to design and manage these kinds of dynamic sociotechnical systems. On the one hand, there is the strong tendency to try to force the system in the direction of the aviation system by restricting the role of the individual user through various sorts of automatic vehicle control systems. In that way, the system would become more controllable in line with traditional engineering norms. On the other hand, there are also small-scale experiments where the task of successfully coordinating all vehicle movements is laid entirely with the road users by deliberately eradicating all the instruments that are customarily used to direct their behaviour, such as road signs, give-way road-marking, traffic lights and all the other paraphernalia. In cases where the global control possibilities of a sociotechnical system are fundamentally limited because for principled or for practical reasons the users of that system are granted a high degree of freedom to involve themselves in the system, it may well be advisable for authorities to resist the temptation to make maximum use of the possibilities for top-down control. Global control does not, of necessity, lead to better results than locally coordinated actions between individual users do.

5.7 CONCLUSION

With the examples given in this chapter, we have shown that traditional engineering opinions about the designing of technical artefacts and about the knowledge that such designing requires is no longer adequate when the artefacts attain a form of complexity that leads us to introduce the notion of sociotechnical systems. The designers and operators of such systems are confronted with numerous aspects that are not easily or not at all describable within the traditional engineering approach, which is overwhelmingly oriented toward the natural sciences. This traditional approach and the accompanying conceptual frameworks, models and theories therefore need to be enriched with knowledge that has been and is being developed within the domain of the social sciences.

As we have seen, one of the features of sociotechnical systems is that they are less predictable than traditional technical artefacts; sociotechnical systems can display unexpected behaviour even if the end-users set out to use the system in a 'neat' or 'tidy' way. Although the notion of a use plan is as problematic for a sociotechnical system as the idea that such systems are designed on the drawing board as a whole, it remains intuitively clear that there are intended and unintended or proper and improper ways of making use of a sociotechnical system. In the next chapter, it will become clear that the presence of emergent properties plays a part in what we can say about the way in which technology develops. Sociotechnical systems also force us to think anew about the way in which we attribute responsibility to designers and users in the ethical questions that surround technology, as will also be borne out in the final chapter.

5.8 A FEW MORE ISSUES

In this chapter, we have argued that the crucial quality of sociotechnical systems is their *hybrid* character. Such systems include both 'hard' technical components and people, but the way in which people fulfil the roles that have been designed for them is fundamentally different from the way a technical component does what it has been designed to do in accordance with the laws of nature. Even though this hybrid character is most apparent in the large-scale infrastructural systems in our society, being large-scale and complex are not prerequisites for having such a hybrid character. Try to imagine the smallest possible system that qualifies as being hybrid. What is minimally required? Can you give an example that meets these requirements? How does it differ most from a purely technical system?

One extremely difficult problem is to describe precisely what a typical sociotechnical system looks like. What are its components? Just 'people' and 'machines', or possibly also other sorts of things? Can, for example, organisations and institutions, which in our society, in their role as 'legal persons', are sometimes treated as being equivalent to persons of flesh and blood, be components of a sociotechnical system? Are the rules and instructions that define the roles of operators perhaps to be treated as components of such systems? But how can abstract things, like rules, and to some degree also institutions, together with tangible things, like people and machines, be components of one and the same thing? But then, is a sociotechnical system itself something tangible? Currently, we lack really satisfactory answers to these questions. Even in the social sciences, where one might expect to find some help, there is, in fact, little agreement on the answers to these kinds of questions.

CHAPTER 6

The Role of Social Factors in Technological Development

The subject of this chapter is the interaction between technology and society but more particularly, between technological development and the context of such development. There are different views about the way in which technology and society influence each other, varying from the idea that technology develops independently of the social context to the diametrically opposed view that the social context determines how technology develops. We shall discuss both these rather extreme viewpoints and provide critical comments before going on to present an alternative point of view in which technological development is seen as the result of what is technically possible and socially desirable. We shall also briefly look at this issue from the perspective of sociotechnical systems as introduced in the previous chapter. First, though, we contemplate why this question occupies such a prominent position in the philosophy of technology. We shall see that it has to do with totally different – positive and negative – views of the role of technology for humans and society.

6.1 UTOPIAN AND DYSTOPIAN VIEWS OF TECHNOLOGICAL DEVELOPMENT

One cannot deny that in the last centuries, technology has advanced tremendously. With civil aviation aeroplane technology, that is certainly the case. In less than a century, civil aviation has turned into the mass transport system we know today. There are various features of that trend of progression that immediately spring to mind. First of all, there is the technological progress that has been made in the designing and producing of aeroplanes; from a technological angle, the achievements of the first civil aviation aircraft of just after the First World War pale into insignificance when compared to the achievements of the present-day planes (see Table 6.1). A second distinguishing feature is the great increase in the types of aircraft that exist; there are now planes for different numbers of passengers, for short and long distance travel, for freight transport, private jets, light aircraft, et cetera. The final notable feature is the massive scale on which aeroplanes are currently deployed. These huge strides in technology have, moreover, not been limited to civil aviation, but occurred equally in terrains such as healthcare, communication technology, car transport and energy provision.

How can these technological developments be explained? What are the driving forces behind such development? Is it primarily a result of increased knowledge and skill or is it predominantly social desires and needs that underlie all of this? Does technology develop autonomously with regard to the social environment or is that development actually determined by society? In trying to answer all these questions, it is important to carefully bear in mind the meaning of the term 'technology' because the answers may depend on how one views technology. In a narrow sense, technology can be seen as technical artefacts (see Chapter 1), in a wider sense, it can be seen as sociotechnical systems (see Chapter 5) and in a still wider sense as being part of society. The meaning of the phrase 'the development of technology' is thus ambiguous so that the question regarding the factors underlying the development of technology must be similarly ambiguous. Depending on the way in which one defines technology, the answer can lean more towards technological factors or, conversely, it can favour social factors more.

Table 6.1: A comparison between the performance of the Fokker F II (1919) and the Airbus A380 (2005).

Aircraft type	Fokker F II	Airbus A380
Length	10.3 m	73 m
Span	17.6 m	79.8 m
Height	3.7 m	24.1 m
Max. number of passengers	4	525
Max. take off weight	2300 kg	560,000 kg
Cruising speed	150 km/u	~ 900 km/u
Cruising altitude	4,000 m	10,000-11,000 m
Max. Range	600 km	15,200 km
Maiden flight	October 1919	April 2005

The problem of the interaction between technology and society and, in particular, the question as to whether it is technology that determines society or the reverse, has for a long time been high on the philosophy of technology agenda. The importance of this issue must be seen against the background of different views of how the development of technology has to be appraised: do the successive advancements in technology contribute to the enhancing of human affluence and well-being or do they lead to exactly the opposite, people's increasing dependence on and subjugation to technology? Both views play a part in Western thinking about technology. A decidedly utopian view of the role of science and technology is given in Francis Bacon's (1561-1626) *New Atlantis*, which is all about a mythical island on which there is a society whose goal is to gain 'knowledge of the causes and secret motions of things; and the enlarging of the bounds of human empire, to the effecting of all things possible'. But also, in more recent times, the praises of technology (and science) are sung. At the other end of the spectrum are the dystopian views where technology is likened to Golem or Frankenstein-like beings, human creations intended to serve the interests of humans but which ultimately turn against their creators. Instead of technology being in the service of people, people are more and more becoming slaves to technology. According to many dystopian views on the matter, technology estranges people from nature and from fellow human beings. Especially in the first decades after the Second World War, the role of technology in society was fiercely debated in conjunction with the possibility of a nuclear holocaust, increasing environmental problems, un-employment caused by far-reaching automation, problems surrounding privacy, et cetera. What is illustrative of the concerns that people had about technological change and the unexpected negative consequences of large-scale technological developments is the fact that this gave rise to a new field of study known as *technology assessment*, the original objective of which was to anticipate or provide *'early warning'* of all the possible consequences of new technological developments.

Against the backdrop of such optimistic and pessimistic attitudes towards technology, the matter of how technology is developing in relation to the social environment suddenly gains special significance. To what degree is it possible to socially influence technological developments, con-sciously contribute to the well-being of people or to ensure that humans are not subjected to the machinations of technology? Given a pessimistic attitude towards technology, the most terrifying doom and gloom scenario one can imagine is that of technology autonomously developing, that is to say independent of any form of human or social intervention. The conviction that we are already living in such a fateful world could be roughly argued along the following lines. Partly due to the ever-increasing level of scientific knowledge there is a constant progress in technological capabili-ties. In addition to that, the lesson learnt from history is that everything that is technically feasible does, in the end, become generally implemented: the mass transportation of people by air, the mon-itoring of people (*'Big Brother'*), the cloning of people and animals, the medical halting of ageing phenomena in humans, and so on. Thus, inevitably, in the wake of the unstoppable advancement of our technological achievements, comes the social application of all the new-found technological possibilities, together with all the negative consequences that it brings. This trend of thought leads to a somewhat fatalistic view of the role of technology in present society, a view that is associated with the notion of 'autonomous technology' described by Winner, L. [1992, p. 13] as 'the belief that somehow technology has gotten out of control and follows its own course, independent of human direction.'

We shall not bother ourselves here with matters about how developments in technology may be assessed, nor with questions relating to scientific and technological policy. Our objective is primarily to gain more insight into how technology develops in relation to its societal (social) context. The first point that immediately needs to be established is that technological development is a very complicated subject which, regardless of the problem mentioned above concerning the ambiguity of the term 'technological development', triggers a whole host of tangled questions. Does it make sense to talk about technology in general? To what extent are views about the factors affecting the development of technology empirically (historically) verifiable? Does autonomous technological development necessarily lead to the far-reaching technological influencing of society? How can the global autonomous technology view be made compatible with local, purposeful technological developments (for instance, in companies)? It will be impossible to deal with most of these problem areas in this chapter. We shall therefore confine ourselves to sketching the most important notions about the relationship between technological developments and their social context. We shall begin with a discussion of *technological determinism*, to which we reckon all views that technology is a decisive determining factor for the development of society.

6.2   TECHNOLOGICAL DETERMINISM

There are many technological determinism variants, and we shall just briefly examine a few of them. One controversial variant, which we shall nevertheless include because of the historical influence it has had, is that of Karl Marx. He it was who famously said: 'The hand-mill gives you society with the feudal lord; the steam-mill, society with the industrial capitalist.' What he meant by this was that if the innovation of new production means leads to new production methods within a society, then the social relations, including economic, political and cultural relations, in that society will also change. Given that technological progress continually leads to new means of production, such technological developments will also inevitably bring shifts in social relations. Because, according to Marx, the ruling classes (the capitalists) remain in control of technology, one can debate whether in this particular case, there is really evidence of a technological determinism variant (in reality, such capitalists determine technological developments and therefore also indirectly how society develops).

The views on modern technology as formulated by Jacques Ellul are pre-eminent instances of technological determinism. According to Ellul, J. [1964], modern technology is now in the phase of totally governing human life. Technology itself has become a new, all-embracing environment from which people cannot escape. That environment is 'of an artificial nature, autonomous, self-determining and independent of any human intervention.' The technical environment (or modern technology) has its own laws. In particular, the technological environment is autonomous with respect to human values and objectives. The only value that counts in the technological environment is that of functional efficiency. A technical artefact is better than others if it fulfils its function more efficiently; no other single value (such as a moral value) is relevant in the evaluating of technical objects. That means that in the technical environment, human values and norms are overshadowed by technological norms. All that counts is the efficiency of means; efficient or more efficient means becomes a goal in itself (which involves an inverting of goals and means). This leads to a situation in which technological rationality dominates: from a technological point of view, the best solution to a social problem is the most efficient one.

Building on the work of Ellul, Winner further elaborated the idea of technological determinism, in particular that of autonomous technology. According to Winner, L. [1992, p. 76], the following two hypotheses might be said to characterise technological determinism: '(1) that the technical base of a society is the fundamental condition affecting all patterns of social existence, and (2) that changes in technology are the single most important source of change in society.' As far as the second hypothesis is concerned, the implementing of new technologies is something that often has unintended and unpredictable social consequences. Because of these uncertain and unintended effects, Winner refers to something known as 'technological drift': we start to drift more and more 'in a sea of unintended consequences' as the speed and large-scale nature of technological development increases. For Winner, modern technology is not simply a collection of implements that we have under control and can use without any danger. For Winner, this interpretation of technology is very misleading and no longer suffices. Instead of being able to freely use technical tools, we are forced to use them in the 'right' way and to ensure that the stipulated conditions for the functioning of the implement are met. In other words, people are expected to adapt to their technological environment; they become slaves of technology instead of technology being their slave. There is evidence of a 'technological order' that is holding the reins of power. Technology lays down the law for us and has thus become a political phenomenon. The question is therefore not so much who is in power but what. The development of technology is not steered by a technological elite (a technocracy); according to Winner, even the technological elite now has no alternative but to bow to the technological imperative and obey it.

The whole idea of technological determinism also forms part of something known as the linear model of science, technology and society (see also in this connection Chapter 4 on the applied-science-view). According to this model, the influencing factors follow a one-directional line going from science to technology and from there to society (see Figure 6.1). Or, to explain it in more detail: fundamental research is followed by applied research, which leads to technological discoveries that, in their turn, give rise to product innovations. Finally, the phases of production and diffusion of such product innovation within society follow. It should be observed that in this model, there is only partial evidence of technological determinism: technology determines society, but technology does not develop itself autonomously since technology does not develop in accordance with its own laws but is instead *grosso modo* an application of science. If one presumes that science develops completely independently of social factors 'in its ivory towers', then one will end up with a model that might be known as 'scientific determinism'.

Figure 6.1: A linear model of science, technology and society.

This, then, concludes our overview of the most important points of discussion with regard to the matter of technological determinism. Following in the footsteps of Bimber, B. [1994], we may distinguish at least three variants of technological determinism, namely the *normative*, the *nomological* and the *unintended consequences* variants.

The normative variant.

This first variant asserts that the technological norms of functionality, effectiveness and productivity take precedence over other value and norm systems, such as ethical and political ones. It is the variant in which technological rationality permeates decision-making in all areas of human life. The normative variant of technological determinism is reflected in the ideas of Ellul and Winner.

The nomological variant.

In its most basic form, this variant is founded on the notion that the future path of technology is exclusively determined by the present situation and by the internal (that is to say, regardless of the social context) developmental laws of technology. Both in Ellul's view of matters and in the linear model of science, technology and society, one can encounter elements that point to nomological variants of technological determinism (in the linear model, the development of technology is deterministic if we presume that science evolves in a deterministic fashion).

The unintended consequences variant.

Not only with regard to technical actions but also with regard to human actions, in general, it is very hard, if not impossible, to predict and control all the consequences of actions. The occurrence of unintended effects cannot be ruled out. When linked to technology, this means that the social consequences of technological developments partly seem to take on an autonomous character from the point of view that they are unpredictable and/or unmanageable for people. This unintended-consequences variant of technological determinism can be identified in Winner's concept of technological drift.

These technological determinism variants are not be confused with each other. They make quite different assertions about how technology develops and about how those developments may be linked to social developments.

A fourth technological determinism variant, namely the *momentum theory* posited by Hughes, T. [1987], is worth mentioning here. This momentum theory focuses on large technological systems. As these systems mature and become more deeply embedded in the social fabric, they develop a type of technological momentum (inertia) which makes them increasingly difficult to influence and to steer; they start to go more and more their own way and to increasingly display deterministic traits. If one is to fully understand Hughes' theory, it is important to realise that, for him, technological systems consist of both technological and social components or, in other words, that he sees technological systems as sociotechnical systems (see Chapter 5). Technological systems can be both causes and results of social changes, but the bigger and more complex they become, the more inclined they are to form society, rather than the other way round. New technological systems that are still young tend to be more open to social influence. This gives rise to the Collingridge's control dilemma (see also Section 7.3). In the phase when goal-directed influencing (steering) of technological systems is still possible, all the unintended and undesired side-effects are often still not known, which means that it is not clear in which direction these technological systems should be steered. By the time those effects have manifested themselves, the technological systems in question have often already reached a phase in which, thanks to their technological momentum, steering is either completely impossible or only marginally possible. Notice that the momentum variant of technological determinism does deviate in one important respect from the previously mentioned variants. According to this theory, technological systems are not *inherently* but rather *de facto* deterministic: in certain circumstances, technological systems go their own way, and in those situations, society seems to have no alternative but to adapt to that path.

A problematic aspect of many technologically deterministic viewpoints resides in the fact that even though technology is a product of human effort – that is to say, the result of human action, – technology itself appears on the scene as an acting agent. If technology is a product of human effort, then it is also a result of human motives, wishes, needs, values, objectives, et cetera. When one examines technological developments from a 'local' perspective (for instance, by looking at technological research programmes within universities or at product development programmes within companies), then one can instantly recognise the influence of such elements. But at a higher level of aggregation, if one takes a 'global' look at technology, then (according, at least, to technical determinists) these elements cannot be detected. Technology has detached itself from human concerns and no longer seems to be the work of humans but rather appears as an autonomous agent, a product of itself that subjugates people and imposes its will. Precisely how these two perspectives can be reconciled tends to remain an unanswered question.

Most of the technological determinism variants leave little or no space for social influence where the development of technology is concerned. Now we shall switch our attention to a point of view where it is precisely social factors that are entirely responsible for determining how technology develops.

6.3   SOCIAL CONSTRUCTIVISM

Social constructivists reject the thesis that technology develops autonomously. According to the theory of *the social construction of technology* (better known as the *SCOT theory*), it is rather people, in particular social groups, that play a crucial part in the development of technology.[32] Technological development is all down to the choices made by social groups and those choices, together with the way in which they influence the development of technology, are not laid down in advance. We shall confine ourselves to presenting the main outlines of social constructivism.

Social constructivists accuse the defenders of technological deterministic ideas that their notions are not based on detailed empirical studies but rather on *a-priori* concepts, such as the distinction between science and technology. They argue in favour of opening the black box of technological development in order to find out what is really happening. They therefore concentrate predominantly on the development of specific technical artefacts and not on the development of technology in

---

[32] See Pinch and Bijker [1987] and Bijker, W. [1995].

general. If you do that, then you will see – so social constructivists argue – that it is social processes that steer the development of a technical artefact. What then emerges is that almost anything is negotiable: whether a certain technical solution works or not, whether somebody may be considered an engineer or a scientist or not, what is certain knowledge and what not, et cetera. The negotiations referred to here take place within the relevant social groups, and those are all the social groups that have a vested interest in a given technical artefact such as the users and the producers but also the designers.

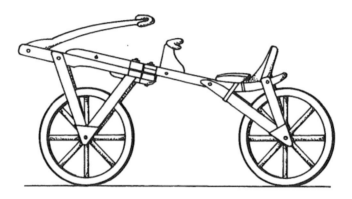

Figure 6.2: The *Dandy Horse* [Van der Plas, R., 1999, p. 11].

Figure 6.3: The *Penny-farthing* [Van der Plas, R., 1999, p. 13].

Figure 6.4: *Lawson's bicyclette* [Bijker, W., 1995, p. 69].

Wiebe Bijker [Pinch and Bijker, 1987] illustrates the importance of social groups with the help of the development of the bicycle. He describes this particular technological development in terms of variation and selection processes. In that way, he sets out to avoid describing the evolution as a straightforward linear process going from the *dandy horse* to the *penny-farthing* and ending with *Lawson's bicyclette*: just as in the field of biology, one cannot speak in technology of directed developments that progress in a straight line. In order to accurately describe the selection process and the success and failure of the various variants of the bicycle, it is necessary to create an historic picture of the way in which the parade of variants was perceived by the relevant social groups of the day. It is those perceptions that determine the significance attached to a certain variant. So it was that sporty young people viewed the penny-farthing in a different light (see Figures 6.2, 6.3, and 6.4) than elderly ladies of the day: the former group saw it as a 'macho' innovation, but the latter group maintained that it was a 'dangerous, indecent' invention. There is thus evidence of interpretive flexibility. In the light of the attributed significance, you can then go on to establish to what degree a variant may be considered to be either a solution or a problem for a particular social group. This therefore means that social groups determine whether a proposed variant 'works', that is to say, whether it amounts to an acceptable solution and, if not, what is problematic about it. A proposed variant can be either more or less acceptable to a certain social group and that, accordingly, will be matched by an either increasing or decreasing stabilisation of a variant within a given social group. If we describe the whole history of the development of the bicycle in this way, then Lawson's bicyclette does not come forward as an end point or temporary end point in the evolution of the bike, as in Figure 6.5, but rather as one of the many variants stabilised by different groups, as in Figure 6.6.

The example of the bike convincingly brings to the fore how important the position of social groups is in the development of technology. According to the SCOT theory, it is a role that goes

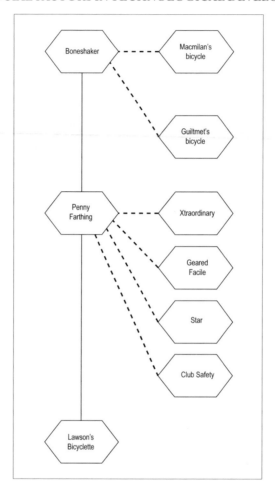

Figure 6.5: "The traditional quasi-linear view of the development of the high-wheeled Ordinary bicycle until Lawson's bicycle. The solid lines indicate successful development, while the dashed lines indicate failure." (caption and figure from Bijker, W. [1995, p. 8]).

much further than has so far been described. We have already mentioned that relevant social groups determine whether or not a technical artefact 'works'. If we adhere to the SCOT theory, the functioning of a technical artefact is also a social construction and that functioning is not determined by characteristics of the technical artefact itself (it is not, for instance, determined by the physical structure of the artefact as we presumed was the case in Chapter 1). By, for instance, redefining various problems, it becomes possible to actually influence the perception of a relevant social group to such a degree that the technical artefact in question does actually work in the end. This means to say

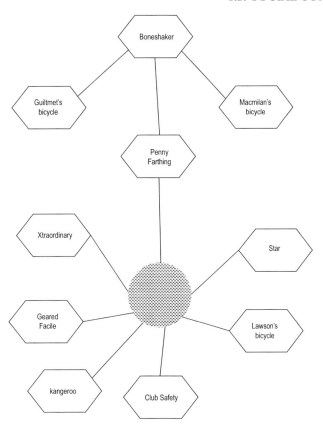

Figure 6.6: "A nonlinear representation of the various bicycle designs since the high-wheeled Ordinary bicycle. The various designs are treated equally, without using hindsight knowledge about which design principles eventually would become most commonly used." (caption and figure from Bijker, W. [1995, p. 9]).

that the great success of one bike variant as opposed to another within a certain social group cannot be explained by seeking recourse to the 'fact' that one bike functions technically better than another. In fact, the impression that one might function better than another is a social construction in itself, a fact which – in its turn – needs to be explained on the basis of social processes (of negotiation).

A problem one can bring in against this view is that it obscures the part played by the physical properties of the bicycle in the perceived good functioning of the bike. One can argue that those properties cease to be of any relevance in social constructivism. Imagine, for argument's sake, the development of a perpetual motion machine of the second kind. One might wonder why it is not, so that among all the relevant groups, there is consensus about the real possibility of such a machine.

Given that it generates net energy and given the importance of the present-day energy problem, perpetual motion machines are most desirable for the relevant social groups to reach agreement about. One of the groups in question consists of physicists, and they argue that it is a fact that such a system amounts to a physical (thermodynamic) impossibility. For social constructivists, though, that physical fact is also a social construction; it is the outcome of all kinds of negotiation processes between physicists and, as such, a social construction. In much the same way that social constructivists do not want to appeal to the physical structure of an artefact to explain how it works and that it works, they do not want to appeal to 'nature' or 'reality' to justify assertions about the impossibility of such a perpetual motion machine. Insofar as concepts like 'nature' and 'reality' have any significance, they are, according to the social constructivists, social constructs. Ultimately, physicists are not prepared to negotiate about the reality of such a perpetual motion machine. From a social constructivist's point of view, such refusal to cooperate seems to be downright morally reprehensible: in view of the fact that we are now consuming our fossil fuels at an increasing speed, everyone should be obliged to contribute to the social construction of such an appliance.

It is not only the working of a technical artefact which, according to SCOT, is a social construction but also the technical artefacts themselves are social constructions. This theory is at the heart of the SCOT approach. What the social constructivists thus mean is that the significance that social groups attribute to an object is what actually turns that same object into a technical artefact. It is those meanings – and the function of the technical artefact is an integral part of that – which constitute the artefact. In their analysis of things, there is nothing other than complexities of meanings. Therefore, when two different social groups attribute different significance to the same physical construction, that means that you are dealing with two different artefacts. This theory thus deviates somewhat from the analysis of technical artefacts presented in Chapter 1. There we, in fact, concluded that technical artefacts have a dual nature; on the one hand, they are constituted by their physical structure and, on the other hand, by their function, and it is the physical structure which guarantees that the technical artefact is able to fulfil its function (that it works). What both theories have in common is that when one and the same physical object has two different functions (meanings), we must be dealing with two different technical artefacts. In line with the dual nature theory that also applies to the physical structure of technical artefacts: when two different physical structures realise the same technical function, then we are dealing with two different technical artefacts. In the SCOT approach, it is all about meanings or significance, that is what constitutes the technical artefact. The physical structure is not of relevance. Apart from that, as we have seen, social constructivists also maintain that the physical features of a technical artefact are the result of social negotiations. Just like the property that a technical artefact works, the physical properties are also social constructions. Even if the physical structure were to constitute part of the technical artefact, the whole technical artefact would still be seen as a social construction.

Social constructivists regard assertions about what is technically and physically possible as matters that are socially constructed. This means that in the final analysis it is only social factors that determine the development of technology. Neither nature nor our technological ingenuity can play a part in the development of technology. The conclusion is, to put it mildly, somewhat diametrically opposed to engineering practice. The designing and developing of technical artefacts is often, in fact, precisely all about 'wrestling' between what is possible and what is desirable. In that struggle, as we shall argue in the next section, both physical-technical and social factors enter into the arena to play their part.

6.4   THE ROLE OF PHYSICAL-TECHNICAL AND SOCIAL FACTORS IN DESIGNING

It is during the design process that a technical artefact is given its concrete form. Changes in technical artefacts in relation to earlier variants are made at that point. When, as in the case of the bicycle, one wants to analyse the factors that play a part in these changes, then it would seem that the design process is the right place for doing that. In this section, we therefore present a schematic framework for analysing the role of physical-technical and social factors in the design process, and thus for analysing changes in the characteristics of technical artefacts [Kroes, P., 1996].   Changes in technical artefacts that occur during production processes and in their social implementation will not be considered here.

In Chapter 2, the core of the design process was described as the establishing of a description of the physical structure of an artefact that can fulfil a given function. The function of the device that is to be designed is normally laid down in a list of functional requirements, which are then converted into a list of specifications or a programme of requirements. That function does not just come out of the blue, it is the result of social negotiation processes in which the various groups involved, including customers but also producers, articulate their wishes and needs. The function of the product that is to be developed is thus a social construction that is based upon what divergent groups consider to be 'desirable'. The aim of the design process is to come up with a description of a physical construction that meets the conditions included in the programme of requirements.

However, not everything that is desirable is also possible, and there can be various reasons for that. The programme of requirements can contain logically contradictory demands: a design parameter like, for example, the weight must be both bigger and smaller than a certain value x. It can also be the case that the programme of requirements conflicts with what is physically possible: for example, information has to be sent at a speed faster than the speed of light. The programme of requirements might also be technically impossible to realise, either because of the technical recourses available to the organisation where the design process is taking place or because of the general level of technology at that particular stage: e.g., there should be millions of electrical connections on a microchip with a diameter of 1 nanometre. It might finally be impossible to meet the programme of requirements because of social, economic or cultural constraints: for instance, in connection with safety requirements, the cost price that has been fixed or due to moral constraining factors. The most interesting case for the analysis of the role of physical-technical and social factors in designing is the case in which a programme of requirements cannot be executed because of the technological level in general. In that situation, the technical preconditions are such that a design that satisfies the whole list of specifications is technically not realisable.

To sum up, both what is desirable and what is possible result in constraints or boundary conditions on the artefact that is to be designed. Together, they determine the solution space for the design problem. As long as no conflict arises between what is desirable and what is possible, it is down to the ingenuity of the designer to think up a solution. When such a conflict does arise, the solution space will, in principle, be empty. What can be done in such a case? That all depends on the nature of the conflict. When a conflict arises between what is desirable and what is physically possible, then the only option would seem to be to redefine what is desirable in the programme of requirements. In the case of a conflict with technical preconditions, that is also a fairly obvious option, but what one can then also do is to try to change the technical preconditions by moving the boundaries of the '*state of the art*' in technology; that can be done by means of starting physical-technical research programmes. The last solution is, of course, only a sensible one if there is a real chance that the limits of what is technologically possible can be changed in such a way that the programme of requirements can ultimately be realised. Whenever there is evidence of a conflict with social, economic or cultural constraints, one can, in principle, try to adjust these constraints by means of social processes of negotiation. Again that would only seem to be a clever option when there is a reasonably chance of success. In cases when such constraints are related to fundamental norms and values in society, then it is very difficult to alter those constraints.

If one represents the artefact that is to be designed as a black box (see Figures 1.2 and 2.1) which, at a given input, is expected to produce a corresponding output, then the whole collection of boundary conditions concerning what is desirable and what is possible can be divided up into two sub collections:

The physical-technical boundary conditions.

The physical and technical boundary conditions have their origin in and are derived from the *content* or future content of the black box. Whichever physical construction is proposed as a solution, it will in any case have to meet the requirement that it is physically possible. But not everything that is physically possible is, at a certain time, also technically possible. That is why the content of the black box will have to meet the requirement that it is technically possible to construct the content thereof.

The contextual boundary conditions.

What falls into the contextual boundary conditions category is all the other conditions. We call those conditions *contextual* because they have their origins in and are derived from the context of the artefact that is to be designed, in which the user's context plays an important part.

One of the most challenging aspects of technical designing and the driving force behind the development of new technical artefacts is the tension which exists between both types of boundary conditions, between what a technical artefact is capable of doing on the basis of what at a given moment in time is technologically possible and the requirements placed upon that technical artefact by the environment (see Figure 6.7).

Figure 6.7: The tension between physical-technical and contextual boundary conditions.

If one takes a closer look at the collection of contextual boundary conditions, then one can observe that they have a very heterogeneous character. They include requirements with respect to the following:

- the primary function of the technical artefact
- the use (safety, user-friendliness, climatological and geographical conditions, et cetera)
- production (safe to produce, mass production possibilities, preconditions derived from production facilities, et cetera)
- maintenance (safety, costs, et cetera)
- market (price)
- insurance
- financing
- legal aspects (patents)
- environment
- technical standards and norms
- aesthetic criteria
- et cetera

What lies hidden behind each of these boundary conditions are the respective interested parties (*stakeholders*); they can be individuals but also organisations, government bodies or social groups.

Whenever there are conflicts between the different contextual boundary conditions, it will be primarily those interested parties that will have to find the solutions (primarily, because one cannot completely exclude the possibility that contextual conditions that might originally have been in conflict with each other, do eventually become compatible thanks to technological innovations; see the discussion on the design of jet engines and noise hindrance in Section 3.4). This would indicate that such contextual boundary conditions are *in principle* negotiable. They are socially constructed in the sense that they are the outcomes of complicated negotiation processes between the various interested parties (this does not mean to say that these negotiation processes can always be controlled; see the discussion on the emergent properties of sociotechnical systems that is given in Chapter 5 and at the end of this chapter). In that respect, it may be said that contextual conditions are essentially different from physical-technical preconditions. Physical boundary conditions are not negotiable on the basis of the interests of the parties concerned. Uncertainties can obviously arise in relation to the validity of physical conditions but such uncertainty can only be dispelled by research, not through negotiations in which various interests are weighed up against each other. We therefore reject the hypothesis put forward by the social constructivists to the effect that in research, physical facts are socially construed. In the case of the technical preconditions, similar considerations might be said to apply. Technological preconditions are not, in principle, negotiable. Whenever there is disagreement about the matter of whether a thing is technologically possible, technological-scientific research will have to provide the answers and not negotiation because in the latter case the outcomes are determined by the interests and positions of power of the parties involved.

There is, therefore, an important discrepancy with regard to the flexibility (negotiability) of physical-technical and contextual boundary conditions. Roughly speaking, it is a difference that can be interpreted as follows. Physical boundary conditions are factual assertions about nature; they have their origins in physically *necessary* relations. They are bare facts in the sense that they cannot be negotiated; they are 'transcendent' boundary conditions in the sense that they lie beyond any human control. Technological preconditions are also factual statements, but they have more of a *contingent* character: they express what in a given historical situation is *de facto* impossible. Constant endeavours are being made to deliberately change the boundaries of what is technologically possible by means of technological research. Those boundaries are not, however, negotiable in the sense that the outcome of the research is determined by the choices of social groups. Finally, there are contextual boundary conditions, which can have both a normative and a factual character. Contextual boundary conditions that have their origins in the desires and needs of interest groups are of a normative character; they are negotiable. If they have their origins in what in a given social situation is possible, then they have a factual character. Insofar as that social situation can, in principle, change (i.e., can be renegotiated), these contextual boundary conditions will likewise be negotiable and flexible. In a certain situation, it is thus possible that in practical terms a contextual condition, like a maximum cost price dictated by the free market, becomes just as 'hard' (or as nonnegotiable) as a physical condition.[33]

---

[33] It is nevertheless thinkable that the social phenomenon of the free market could become a political point of debate, so that, for instance, via subsidies, boundary conditions with respect to cost price could be renegotiated. For physical and technical conditions, such a scenario is unthinkable; we cannot negotiate ourselves in a different kind of physical reality, even though we might want to.

In contrast to the theory of social constructivism, we see that in the framework presented here, both physical-technical and social factors each play their own part in the realisation and development of technical artefacts. It is precisely the tension between what is technically possible and what is (socially) desirable, that, to a large degree, determines the ultimate form and thus the actual development of technical artefacts. According to our framework, there is roughly speaking evidence of a certain asymmetry in the flexibility of physical-technical and contextual boundary conditions. That has important consequences in situations where conflicts arise between both kinds of boundary conditions. In view of the fact that only the latter are, in principle, renegotiable, it follows that in such situations there is no alternative than simply to adjust the contextual boundary conditions.

From a social-constructivist point of view, one can rebut that the demarcation line between physical-technical factors and social factors is again negotiable. Yet, even if this point is accepted, it does not show that the distinction between physical-technical and social factors does not hold. One can have distinctions which allow seamless transitions and that nevertheless single out significant differences, just like we saw in Chapter 1 for the distinctions between technical artefacts and natural objects, and between technical artefacts and social objects. In connection with this rebuttal, a final remark needs to be made. We have presented our framework as if the domains of that what is possible and of that what is desirable, are completely divorced from each other. Naturally, that is not the case: what is desirable is to a large degree conditioned by what is technically feasible and, conversely, the boundaries of what is technically feasible are perpetually forced to shift under the pressure of what is desired. The two poles between which the development of technical artefacts takes place are thus not completely independent of each other.

6.5   THE EVOLUTION OF CIVIL AVIATION

We shall now return to the example of the development of civil aviation. We began by describing three features of that development. The first was the technological progress involved in the design and production of aeroplanes. It is tempting, when considering the tension between what is desirable and what is possible, to opt for what is technologically possible as the *prime mover* of progress in aviation technology. If one also assumes that what is technologically possible is a direct consequence of an autonomous kind of progress in the natural sciences, then one is quickly back to a linear interpretation of technological development. However, it would be naïve to maintain that it was solely or especially the autonomous advancement in the natural sciences that facilitated all the developments in the field of aviation technology. Not very long after it had been invented, it was realised that the aeroplane was of great military-strategic importance, which is why the military sector may be said to have had a large influence on these developments. That, in turn, had an effect on the development of civil aviation. It is undeniable that the wishes and needs of the industrial-military sector have, generally, deeply influenced the way in which the natural sciences and technology have developed in the twentieth century. But it is also true to say that more or less autonomous developments in the areas of natural scientific and technological knowledge have produced completely unexpected applications that have also been of importance to the progress of aviation technology. Such development has thus been the result of both internal developments in science and technology and contextual (social) factors.

The second feature of the development of civil aviation was the huge increase in the diversity of aircraft types. Such innovations can be partly linked to new technological possibilities (consider, for instance, the difference between propeller and jet propulsion) but also to the desires and needs of customers. The first link brings us back to the previous point about how to explain the development of new technological possibilities in aviation technology. The second link seems, at first sight, a predominantly contextual factor at work. However, that is only partly true. When, just like in the example of the bicycle, we explain the range of aircraft types on the basis of variation and selection processes, then the role of physical-technical factors comes clearly to the fore. As we have argued, when it comes to the variation of aeroplanes in the design and development phase, then there are both physical-technical and contextual boundary conditions that play a part. In the selection processes, by contrast, it would appear to be predominantly contextual conditions – the wishes and needs of customers or of more generally relevant social groups – that come into play.

The final feature we drew attention to was the vast scale of the deployment of aeroplanes for the transportation of people and freight. This facet of the developments in civil aviation would seem to be difficult to explain on the basis of developments in terms of technical preconditions or on the basis of autonomous technological development. Perhaps a fanatical defender of the autonomy of technology might claim that the tendency towards ever bigger aircraft may be explained on the basis of considerations concerning energy efficiency or aerodynamic considerations, but that would hardly explain the vast scale of the deployment of aeroplanes. Here, the balance seems to tip clearly in the direction of contextual factors. Nonetheless, one should be aware that important technological developments (like modern air traffic control systems and mass production technology) have also contributed to making these developments possible. So once again, despite the dominant role of social factors, technological aspects are very much in the picture.

6.6   CONCLUSION

Both the idea of autonomous technology and the theory of social constructivism are inadequate when it comes to explaining the developments in civil aviation. Both merely provide a partial view of these developments. In reality, there is always evidence of a complicated interplay between physical-technical and contextual (social) factors; in that, interplay attuning takes place between what is possible and what is desired.

Does nothing then remain of the notion of autonomous technology that is dominating society? Can one, on the basis of the analysis proposed in this chapter, conclude that all the worries of many philosophers of technology have been unfounded? Definitely not. There is undoubtedly a kernel of truth in the normative and unintended consequences variants of technical determinism. With the ever-increasing importance of technology in our society, it would appear that thinking in terms of technological rationality (in terms of more efficient means), is gradually getting the upper hand. Furthermore, it cannot be denied that because of the large-scale applications of technology humankind is being confronted with unintended consequences, global warming being merely the most recent example of that. As we saw in Chapter 5, many technological systems do, in fact, form an integral part of more extensive sociotechnical systems in which social structures and people also have a role to play. Such sociotechnical systems often display emergent behaviour, behaviour that is difficult to predict, despite the fact that it is the result of the purposeful dealings of the many different operators within the system and users of the system. That unpredictable behaviour is by definition unintended. It would thus seem that, at the systems level, sociotechnical systems behave autonomously. The hypothesis on the autonomy of technology attributes the autonomous behaviour of sociotechnical systems to one component of sociotechnical systems, namely to the technical subsystem. But that is incorrect. Technology does not develop autonomously but sociotechnical systems do, to a certain extent, develop autonomously.

6.7   A FEW MORE ISSUES

In this chapter, we have implicitly presumed that technology development is also progress, and we have asked which factors are responsible for that progress. But is there evidence of progress in technology? Can you give examples of technological knowledge and technological skill that has been lost over the course of time? How would you, or perhaps could you, define technological progress? Does technological progress also mean that there is progress in the social implementation of technology? Is it, indeed, the case, as put forward in the doom and gloom scenario, that everything that is technologically possible is also socially implemented? Can you give examples of this?

Having insight into the factors that are responsible for technological development does not automatically mean that it can be controlled or steered on the basis of such insight. Do you think that it is possible to influence technological development? If not, why not? If you answer negatively, does that automatically mean that you are a convinced technological determinist, or could there be other reasons for thinking that technology cannot be influenced? If you think that technological development can be influenced, then how can technological development be influenced or steered? Should one think primarily of the steering of applied scientific research or of steering through legislation and rules for technological development within the free market business sector? Or are there other possible steering mechanisms?

CHAPTER 7

Ethics and Unintended Consequences of Technology

In Chapter 5, we have seen that technical artefacts are often part of larger sociotechnical systems and that those systems also contribute to determining the consequences of the use of such artefacts. We ended Chapter 6 with the conclusion that technology can result in unintended consequences. These two observations lead one to wonder to what extent engineers are able to predict the consequences of what they design during the actual design process. Those consequences provided an important basis for the ethical questions accompanying the design processes that were dealt with in Chapter 3. What happens if these consequences are not known or are perhaps not foreseeable? What kind of ethical questions does that then precipitate? How can and must an engineer cope with such matters?

7.1 UNINTENDED CONSEQUENCES AND RISKS

Technologies have unintended effects. Aeroplanes crash with fatal consequences. Chemical plants pollute the environment. Security cameras diminish privacy. In other words, such technologies can be dangerous for humans, the natural environment and animals.

The various dangers and other unintended consequences are often not known beforehand, but that does not always mean to say that they are impossible to know. One strategy towards the unintended consequences and dangers of technologies, and addressing them in time, is therefore to try to predict the consequences as good as possible and to expressing the dangers in terms of risks. What are the possibilities and constraints of such a strategy?

'Risk' is usually defined as *the probability of an undesired event* times *the consequences of that event*. If we want to express dangers of a technology in terms of risks, then we need to have reliable knowledge about what precisely the consequences of the technology can be, and we also need to know what the probabilities are that the consequences might materialise. For technologies that have been in use for a long time and where a number of incidents and accidents have occurred, we tend to have reliable information about both factors in the form of accident statistics. We are, for instance, able to estimate the probability of a passenger dying in an air crash in terms of the probability of being killed per kilometre flown. Much the same applies to most other forms of transport.

In the case of new innovations, such reliable information for calculating risk levels is usually lacking. We are often not aware of all the unintended consequences and dangers of a certain new technology. And even if we are aware of the possible undesired consequences, we do not always have enough knowledge about the failure modes: the possible ways in which a new kind of technology can fail. Apart from anything else, in the case of new and innovative technologies, we do not have accident statistics for calculating failure probabilities for the simple reason that no accidents have yet occurred. In such cases, engineers often employ fault trees or event trees in order to estimate the probability of failure. An event tree sketches possible sequences of events that can follow some kind of potential technical failure, like the failure of a plane's landing gear to properly operate. A fault tree sketches the possible series of events that can lead to an accident such as, for instance, the crashing of an aeroplane. By attributing probabilities to individual events in an event or fault tree, the probability of certain accidents can be calculated. Generally, speaking, predicting risks in such a way tends to be less reliable than making use of real accident statistics.

Let us now presume that despite these constraints, we are able to generate more or less reliable estimates of the risks of a certain technology. What then is the use of such risk assessments? A possible way in which we could use such risk assessments is to judge their acceptability. Indeed, the smaller the risk, the sooner we will be inclined to decide that the advantages outweigh the risks of the technology in question.

Although the magnitude of its risk says something about the acceptability of a technology, this magnitude does not tell the whole story. Also other considerations are relevant. To see this, suppose that you want to compare the acceptability of two very different types of technologies. Some engineers and scientists maintain that if according to given risk assessments the risks of these two technologies are roughly the same, then those technologies are also morally equally acceptable. This would, for instance, mean that if you accept one technology, then you would be obliged to also accept the other one with approximately the same risk level. However, this line of reasoning is incorrect for the following reasons. In the first place, not all risk assessments are equally reliable. It is therefore not always appropriate to compare the outcomes of different risk assessments with each other. In the second place, risk assessments are often multi-dimensional whilst the risk comparisons are often one-dimensional, or at least confined to just a few aspects. A very commonly used risk measure is that of the number of fatalities per unit of time. In reality, though, two technologies that are equally dangerous in terms of the number of fatalities per unit of time, can be very different in terms of the danger they pose to human health, the damage to the ecosystem, their economic threat, et cetera. In the third place, the acceptability of risk does not only depend on how big a risk is but also on the question of whether the risk is voluntary and whether people have agreed to take the risk. The risks posed to people that take part in traffic thus tend to be more voluntary accepted than, for instance, the risks posed by a chemical plant near where they live, especially in view of the fact that chemical plants are frequently just built without first consulting the local inhabitants. On the other hand, the risks attached to skiing, for example, are more voluntary than those attached to participating in the road traffic system. On the whole, voluntarily accepted risks are seen as being more morally acceptable than those that are forced upon people. In the fourth place, risks are not just inherently acceptable but are acceptable because they bring with them some kinds of advantages. That means that if the advantages are great, people have reason to accept higher levels of risk. In the fifth place, the acceptability of risks also depends on the availability of alternatives. If there is a possible alternative that carries fewer risks and has no other major disadvantages, then the very existence of that alternative might be grounds enough for viewing the risks of an existing type of technology as undesirable. Finally, the acceptability of a given risk depends on how fairly the risks and benefits are distributed. If there is a certain group in society that only has the disadvantages of the risks without being able to enjoy any of the benefits attached to such risky activities, then that can also make the risk in question morally unacceptable.

In the discussion provided above, there are four important considerations that come to the fore in judging risk acceptability. The first is the question of whether the advantages of the risk-bearing activity outweigh its disadvantages. The second is whether there are alternatives that would, in that respect, score better. The third consideration is that of whether the risk is voluntarily taken and whether or not those involved have agreed to the risk. In ethics, this agreement is often described as giving informed consent. The main idea is that a risk is only acceptable if individuals agree to it after having first been fully informed of the risks, advantages and technological alternatives. A fourth matter of consideration is whether all the possible advantages and disadvantages of the activity-bearing risks are fairly or justly distributed. The value at stake here is that of distributive justice.

Risk assessment may thus be useful when assessing the magnitude of risks. A second reason for possibly wanting to make use of risk assessment might be to limit the risk magnitude at the actual design stage. This is often viewed as an important moral responsibility of engineers. Many professional codes of conduct emphasise that engineers are responsible for designing safe installations. As is stated in the code of conduct of the NSPE, the National Society of Professional Engineers in the USA:[34]

Engineers shall hold paramount the safety, health, and welfare of the public.

Although it would generally seem desirable to minimise the risks during the design process, it is not always possible or, for that matter, desirable to do so. It is not always possible because there is no such thing as artefacts or processes that are absolutely safe. It is not always desirable because reducing a certain risk often brings with it costs or other kinds of disadvantages. For example, a safe car or aeroplane is usually more expensive. If planes are made safer by being made heavier, that means that they will use more fuel which, in turn, will make flying more expensive and will further encumber the environment. In these sorts of cases the pros and cons have to be morally weighed up in order to determine whether it is worth reducing the risks. That is where the above-mentioned considerations play an important part.

If it is decided that the risks should be limited during the design process, then there are different strategies that can be adopted. Following Sven Ove [Hansson, S., 2007], we shall discuss four of those possible strategies.

A first strategy is *inherently safe designing*. This strategy aims at taking away dangers rather than managing them. That can, for instance, be done by substituting various dangerous substances, mechanisms or chemical reactions with less dangerous ones. In order to do that, one needs to have

---

[34]NSPE Code of ethics, from `http://www.nspe.org/Ethics/CodeofEthics/index.html` Accessed October, 19 2009.

knowledge of the kinds of substances, mechanisms or reactions that constitute a danger. It is not, however, necessary to have details concerning the exact probabilities of the dangers arising. To follow this strategy you need to have some but not full knowledge about the risks of a given technology.

A second strategy that can be adopted is that of building in *safety factors*. This means creating a structure so that its expected strength exceeds the expected load with a certain factor, the safety factor. The risk of failure of that particular construction is thus reduced. It is also a way of coping with the possible uncertainties attached to the expected loads and the predicted strength of the construction. Indeed, there is a tendency to think more in terms of the expected probability of failure of a construction or of components – that may not exceed certain thresholds – than in terms of the safety factors since the former results in less over-dimensioning and thus cheaper constructions. This, however, requires reliable knowledge of both the expected loads and the actual strength of the construction, and it eliminates uncertainty margins which are incorporated in a design if it is based on safety factors.

A third strategy is that of incorporating *negative feedback*, which involves designing installations in such a way that if an installation fails or if a operator loses control, a mechanism will be activated that will automatically switch off the installation. A good example of this is the driver's safety device (DSD) in trains which ensures that the train comes to a standstill if the driver falls asleep or loses consciousness. Note that up to a certain point, this strategy can cover partially unknown risks: a mechanism can be triggered if, for instance, a certain poisonous substance is in danger of escaping from a chemical plant. One does not need to know beforehand what the probabilities of such an incident occurring are in order to be able to design an effective negative feedback mechanism.

A final strategy can be to design *multiple independent safety barriers*. In that way, one creates a chain of different independent safety barriers to ensure that if the first barrier fails, there will be others to back it up. In this way, the probability of danger arising is reduced. With this strategy, it is once again the case that one does not need to know beforehand what precisely the probabilities of an incident occurring are or what the exact causes might be. In order to be able to design safety barriers that are actually independent, one really needs to have insight into the possible failure modes; otherwise, there is always the danger of a failure mode simultaneously undermining different safety barriers.

This brief overview of strategies shows that during designing, one can take measures to reduce possible dangers, even if one is not exactly aware of the dangers and one is not able to precisely quantify them in terms of risks. All the same, one must have some idea of what can go wrong and what unintended effects can occur if one is to successfully apply the referred to strategies. In other words, it is more or less impossible, during the design process, to account for unintended consequences which cannot possibly be foreseen beforehand. The unintended consequences variant of technological determinism discussed in Section 6.2 posits that such unintended and unforeseen consequences cannot be avoided in technological development. We concluded in the previous chapter that this assumption regarding the unintended consequences variant within technological determinism is indeed plausible. The conclusion must therefore be that during the design process, one cannot take into account *all* the possible unintended consequences of a technology because some of the consequences do not reveal themselves beforehand.

7.2   SOCIOTECHNICAL SYSTEMS AND RESPONSIBILITY

One reason why one cannot account for all the consequences of technology beforehand is because there is a limit to the amount of time that can be spent on finding out about those consequences. Quite where one places that limit is also an ethical question because the decision to finishing a design process quickly, may have the consequence that the probability increases that users and others are burdened with the negative consequences of technology.

There is, however, another reason why one cannot always predict the unintended consequences of technology and that is this: during the design phase the consequences are indeterminate because they also partly depend on the actions of other *actors* besides the designers, like the users. This especially applies when we take sociotechnical systems into account, which – as we saw in Chapter 5 – depend for their proper functioning on all sorts of actors, such as operators.

If undesired consequences arise in a sociotechnical system it is, in many cases, not possible to simply trace the cause of those consequences back to one actor who might have been able to foresee and prevent such consequences and who can, thus, be held responsible for them. It is much more likely that the consequences will depend on the actions of a number of actors and on the constellation of the sociotechnical system as a whole. This was a point that clearly emerged from the discussion in Chapter 5 on the mid-air collision above Überlingen (see Section 5.5). The final upshot would appear to be that one often cannot indicate who is responsible for certain undesired consequences. This is also sometimes known as *the problem of many hands*: because there are so many people who play a small (and difficult to pinpoint) part in the chain of the events that it is difficult to establish who is responsible for what.[35]

The problem of many hands is not just a problem because, in retrospect, no one can be held responsible but also because, apparently, no one feels the obligation to endeavour to prevent such consequences occurring, and so we do not learn from our mistakes. All the same, there are times when it is possible to hold one or more individuals responsible for an accident. To illustrate this point, we shall now examine an aeroplane accident that took place in 2007, in Brazil.

On 17th July 2007, an Airbus A320 overshot the end of the runway when landing at Congonhas International Airport in São Paulo. It ploughed across a motorway and finally came to rest, next to a petrol station, in a warehouse belonging to the Brazilian airline company TAM. It then exploded. Some 199 people were killed in the incident, 12 of whom were on the ground.[36]

The initial conclusion was that the accident was attributable to the relatively short runway and to wet weather. The runway in question had been resurfaced not long before, which meant that gullies had not yet been created to deal with excess water on the runway and thus prevent *hydroplaning*. Already airport safety there had previously been questioned, especially in wet weather.

---

[35]Bovens, M. [1998].
[36]http://en.wikipedia.org/wiki/TAM_Linhas_A%C3%A9reas_Flight_3054, last visited on 7th December 2007.

In February 2007, a Brazilian judge had imposed a flying ban at Congonhas for a number of aeroplane types, following complaints lodged by pilots about rainwater on the runways, which had negatively influenced the planes' breaking performance. The airport, the Brazilian airline company TAM and the Brazilian Civil Aviation Authority all contested the ban, and so within a day it was dropped following a higher court overruling. The A320 did not fall under the original ban because according to Airbus, its braking distance was shorter than that of the banned aircraft types.

When it became clear that during the landing of the fatal flight water accumulation on the runway was limited, the investigation focused on other possible causes. What soon emerged was that the thrust reverser of the right engine had not operated during the landing procedure. Thrust reversal is deployed during touchdown to aid deceleration. The authorities knew about the problem, but according to Airbus and TAM, the airline company, it was not unsafe to land with a defective thrust reverser. According to reports – that were vehemently repudiated by TAM – there had been landing problems with the same plane only a day before.

From analyses of the *Flight Data Controller* and the cockpit conversations, it became evident that the pilots had been aware of the problem with the inoperative thrust reverser in the right engine. During the landing roll, they had switched the right engine to 'climb', probably to prevent the defective reverser from starting to operate. Because of that, the spoilers on the wing did not work. During landing, those spoilers are used to increase the plane's air resistance or drag while at the same time reducing the lift factor so that the plane is forced down onto the runway and can brake more easily. With the Airbus A320, the *spoilers* only work properly if the engines are in the 'idle' position, but as the right engine was in the climb mode during that landing procedure, the spoilers did not come into play. Most probably, that was the cause of the accident or, at least, one of the major contributing factors. Interestingly, after the disaster, Airbus ordered that with that particular series of aircraft, all engines should be switched to the idle mode in readiness for touchdown.

Who, then, was responsible for the Brazilian crash? Before answering this question, it is useful to first think about the conditions under which someone can be held responsible for something. Over the course of time, philosophers have given various answers to this very question. In many philosophical discussions, there are a number of conditions that recurrently arise. The following conditions are the ones that are most frequently mentioned: [37]

1. The action taken by somebody for which (s)he is held responsible must have been consciously undertaken;

2. There must be a causal connection between the action taken and the ensuing consequences for which someone is held responsible;

3. The person must have foreseen or at least have been able to foresee the consequences;

4. The person could have taken a different course of action;

5. The action taken was wrong or otherwise blameworthy, and that mistake or blameworthiness contributed to the negative consequences.

[37] See for example Bovens, M. [1998] and Fischer and Ravizza [1993].

In the light of this information, look once again at the actions of the pilots. The relevant actions of the pilots involved consciously adjusting the right engine to the 'climb' mode or, more precisely, leaving it in that position. In so doing, they fulfil the first condition. Thanks to this course of action, though, the spoilers did not function in the way that they should have done, so the plane failed to decelerate sufficiently, flew off the runway, crashed and went up in flames. Condition 2, concerning the existence of a causal connection, is also satisfied. Whether or not the pilots could have foreseen the disaster, which is the third condition, is less clear. One presumes that the pilots were not able to anticipate the consequences of their actions: they did not deliberately crash the plane. The real question is whether they could have or should have been able to anticipate the consequences. One argument that can be levelled is that according to the flying instructions for that particular type of Airbus, the engines should be in the 'idle' mode when landing; the onboard computer also backed up those instructions, but the pilots chose not to follow them. It could be argued that pilots should know that failing to follow the flying instructions or to observe the directions of the onboard computer can lead to a disaster. The question as to whether the pilots could have done any differently, the fourth condition, is disputable. At first sight, it would seem obvious that the pilots could have done something differently: after all, they could have allowed the right engine to idle instead of climb. Nevertheless, one could argue that the pilots were forced to make a quick decision in what was not an everyday situation (with a malfunctioning thrust reverser in the right engine), and that it was perhaps not obvious that it would make a difference in that situation if the right engine was in the climb mode. On the other hand, one could argue that it is part of the pilot's job to be able to react quickly and appropriately in such situations. As far as the fifth condition is concerned, the pilots clearly breached a norm, which means that their actions could be labelled wrong. They failed to abide by the flying instructions given in the manual (see also Figure 7.1). On top of everything else, they ignored the directions given by the onboard computer. Still, one can argue that this was an unusual situation in which the pilots had to think and react fast, so they could not be blamed for what they did wrong.

Apart from looking at the pilots, one can also look at other actors that could possibly have borne some responsibility in this case such as the air traffic controllers, the aviation authorities, the plane's maintenance technicians and the engineers, who designed the aeroplane and drew up the flying instructions. Just think for a moment of the engineers and go once again down the list of the five conditions.

One may presume that the engineers did make the relevant technical decisions in a conscious fashion; they knew what they were doing or should at least have been aware of what they were doing. They therefore satisfied the first condition of responsibility. To what extent the second condition was also met – that of a causal connection between the actions of the engineers and the disaster – is not so clear. If the design had been different, the disaster may possibly have been averted. In addition, the manual could have stated that the aeroplane should only be allowed to take off if both thrust

---

[38] *Flights Control, Flight Crew Operating Manual, Airbus A 320.* Available at `http://www.smartcockpit.com/data/pdfs/ plane/airbus/A320/systems/A320-Flight_Controls.pdf`, last visited on 23rd June 2008.

AIRBUS TRAINING A320 SIMULATOR FLIGHT CREW OPERATING MANUAL	FLIGHT CONTROLS	1.27.10	P 12
	DESCRIPTION	SEQ 001	REV 37

GROUND SPOILER CONTROL

Spoilers 1 to 5 act as ground spoilers.
When a ground spoiler surface on one wing fails, the symmetric one on the other wing is inhibited.

Arming
The pilot arms the ground spoilers by pulling the speedbrake control lever up into the armed position.

Full extension
The ground spoilers automatically extend during rejected takeoff, at a speed greater than 72 knots, or at landing when both main landing gears have touched down, when :
R   · Ground spoilers are armed and all thrust levers are at or near idle, or
R   · Reverse is selected on at least one engine (other thrust lever at or near idle), if ground spoilers were not armed.

Figure 7.1: A section of the '*flight crew operating manual*' for the A320.[38]

reversers are operating properly. That would probably have been sufficient to prevent the accident. To conclude, it would therefore seem that the engineers' manuals did have a causal connection with the accident. To what extent the engineers could have foreseen this – the third condition – is once more not so clear. When designing the aircraft, the engineers probably did not take into consideration a scenario like that which occurred during the Brazilian air disaster. One can really question whether the engineers should perhaps have borne in mind the possibility of something like that happening.

In this particular case, it would seem that the fourth condition was met: the engineers probably could have dealt with matters differently. There are, at any rate, no indications that they were forced to adhere to this design or were somehow pressured to do so by their superiors. The mention, in the flight manual, that aeroplanes can take off if one of the thrust reversers is out of order is probably dominated by commercial considerations because it means that a plane can put in more flying hours before requiring maintenance.

The fifth condition would imply that the actions of the engineers might have been incorrect or blameworthy. In this connection, the following stipulation laid down in the NSPE code of conduct is of relevance[39]:

(II.1.b) Engineers shall approve only those engineering documents that are in conformity with applicable standards.

As far as it is known, the aeroplane did satisfy all the safety regulations. It is possible, though, that the design endangered the well-being of society in a different way. In this light, the following

[39]NSPE. 2010. *NSPE Code of Ethics for Engineers*. National Society of Professional Engineers, USA 2007 [cited 10 September 2010]. Available from http://www.nspe.org/Ethics/CodeofEthics/index.html.

stipulations in the professional code of conduct of the Dutch association of engineers KIVI-NIRIA are relevant[40]:

> *(1.3.1) The engineer must carefully evaluate the safety and reliability of the systems that have been designed and for which he is responsible before going on to give his approval.*

> *(1.3.2) The engineer must provide manuals (containing the relevant standards and quality norms) so that the user is able to make safe and correct use of the products and systems for which the engineer is responsible.*

Although there are no clear indications that the engineers violated any of those stipulations, there are a number of questions that could be asked. One might, for instance, query whether the engineers had sufficiently evaluated the safety of the system. As has already been suggested, one could ask whether the engineers might not have been able to anticipate events of the kind that took place in Brazil in 2007. If they had, then they probably would have designed the system slightly different. Regarding the second directive, directions for use had been drawn up that would also have made it possible for the pilot to land safely in these circumstances. As was suggested above, one might ask whether those instructions for use were not too complex; the pilots had to react in a very short space of time. From the point of view of safety, the manual should perhaps also have mentioned that planes should only be allowed to take off if both thrust reversers are in working order. Finally, one might wonder to what extent the engineers were involved in or were able to influence this particular aspect of the instructions for use.

There are therefore arguments for holding responsible both the pilots and the engineers who designed the aeroplane. However, there are also arguments against holding these particular actors responsible. It has to be acknowledged that there were many other actors involved who, directly or indirectly, may have contributed to the disaster. For instance, the Brazilian Civil Aviation Authority had been held responsible for previous accidents; some even spoke of a safety crisis within the Brazilian aviation sector. There had also been complaints about lack of safety at the airport where the accident happened. One might even go as far as to assert that the cause of the crash needed to be sought in the sociotechnical system as a whole. This was definitely so in the case of the accident discussed in Chapter 5 where there was a mid-air collision. In these kinds of situations, the matter of responsibility is more diffused. As we also saw in Chapter 5, it is not usually one actor that designs a given sociotechnical system; the system is much more often the – partly unintended – result of the actions of many actors.

We may therefore conclude that if engineers design components or artefacts that are destined to operate within a wider sociotechnical system, then the social consequences of those components or artefacts are hard to anticipate in the design phase because they partly depend on the actions of other actors and the character of the sociotechnical system. It would, however, be too simple to conclude that, in this case, the responsibility for the possible negative consequences should solely be

---

[40]http://www.kiviniria.net/CM/PAG000002804/Gedragscode_2006.html, last visited on 23rd June 2008. Our translation.

placed on other actors. We have seen that in the case of the aeroplane incident in Brazil, it is certainly possible to argue that the engineers also bear a certain amount of responsibility. We also saw that in sociotechnical systems, in general, it is often not easy to identify who is responsible for what. That could lead one to surmise that no one is responsible. Yet that would seem to be a rather unattractive outcome, given the widely supported desire to develop technology in a socially responsible way, and in the light of the knowledge that if all concerned attune their actions to those of others, it may well be possible to prevent certain negative consequences. In the next section, we shall therefore consider how an engineer can contribute to responsible technological development provided that (s)he recognises that technology sometimes has unintended and undesired consequences that could not have been fully foreseen in the design phase and provided that (s)he recognises that those consequences also depend on other actors as well as the sociotechnical system within which the artefacts operate.

7.3   TECHNOLOGY AS A SOCIAL EXPERIMENT

In the light of the above-mentioned findings, one might wonder if it is a good idea to address ethical questions already during the design process in the way suggested in Chapter 3. Might it not be better to wait until the consequences become clearer? We would then, at least, know what we are talking about.

Letting the consequences of technology manifest themselves after the designing of a technology, also has a number of disadvantages. It is then not only the case that these undesired consequences will arise, but also the costs of preventing the consequences from arising again in the future will often be much higher than when matters are tackled in the design phase. Returning to the drawing board is not only expensive because something new has to be designed but also because – once the consequences manifest themselves – the relevant technology will already have become embedded in society. That means that users will already have become used to it and will have adapted their behaviour to that technology; it also means that regulations and other social institutions will have become adjusted to that particular technology. Breaking open such a level of embeddedness is often not only difficult (if indeed possible at all) but also expensive. Furthermore, as we saw in the last section, if negative effects are manifested in a sociotechnical system, it is often not very easy to determine who exactly is responsible for those effects. That, in turn, can lead to situations in which people avoid to address those undesired effects once they have manifested themselves.

When dealing with the possible unknown effects of technology, we are therefore confronted with a dilemma. On the one hand, we often do not know, in the design phase of a given type of technology, the possible future consequences that we have to take into account when designing. On the other hand, by the time those consequences become manifest, the technology in question is already operational and the costs of redesigning or making other adaptations within the sociotechnical system are often high and difficult to realise. This, in fact, is the Collingridge's control dilemma that was already described in Chapter 6.

Although there is no easy solution to the Collingridge's control dilemma, one might not conclude from the mere existence of the dilemma that responsible technological development during the design phase is impossible. In fact, in many cases one does, already in the design phase, really have some knowledge of how a certain kind of technology will probably be implemented and of the sort of consequences which that will have. The ideas discussed in Chapter 3, relating to designing responsibly, such as the notion of value sensitive designing, have not therefore suddenly become nonsensical. The analysis given in this chapter shows that those ideas are not enough to prevent all the possible negative consequences. For example, when implementing a certain technology, it might become apparent that a value sensitive design does not work, or that in practice it gives rise to a new value conflict. In a certain respect, we already noticed that in the example discussed in Chapter 3 on the development of aeroplane engines. Despite the fact that it has become easier to develop ever-more quiet aeroplane engines, as far as society is concerned the noise factor problem has only been resolved to a limited extent because flight movements are increasing. This increase is a development that has been partly made possible by the creation of quieter engines: it has meant that noise hindrance has become less of an issue and so the aviation sector and fleets have been able to expand. We thus see that technologies that seem, on the drawing board, to be capable of resolving certain problems do not always manage to do that in practice. This obviously does not mean that one should therefore stop designing in a value sensitive way; it means that one must simply be aware of the boundaries and limitations of such an approach.

To a certain extent, it is probably possible, during the design phase, to anticipate the fact that the consequences of technology will be partly unknown or undetermined. This can, for example, be achieved by endeavouring to come up with designs that are robust, flexible and transparent.[41]

A design may be said to be *robust* if it functions well in different and preferably also unforeseen circumstances. The strategies discussed in Section 7.1, for making designs more resilient to risks and dangers contribute to robust designing. There we saw that up to a certain point such strategies are also suitable for dealing with unknown factors. Another possible way of making a design robust is by building in redundancy, for instance, by creating certain vital components in duplicate so that if one fails the other can take over. A sociotechnical system can, for example, be made robust by ensuring that if one component fails, it does not immediately bring down the entire system.

A design can be *flexible* in different ways. It can be flexible in a physical sense, which means that it can relatively easily be adapted or rebuilt. In this respect, many modern appliances are either not very flexible, or they are not flexible at all; for instance, they often cannot be opened so that a component part can be replaced. A design can also be termed flexible in a functional sense: it can fulfil different functions. Likewise, sociotechnical systems can display greater or lesser degrees of flexibility; not only in the sense that they can fulfil different functions but also from the point of view that the various component parts can work together in different ways to arrive at a system that functions well. Flexibility makes it possible to react and adapt to unexpected eventualities. Finally, in a flexible system, it is generally less expensive to make adjustments to dispel the undesired consequences that only manifest themselves during use or implementation.

---

[41]For similar and other suggestions, see Collingridge, D. [1992].

Transparency means to say that a technical artefact or a sociotechnical system and the way in which it works, is clearly understandable to the users and, in the case of sociotechnical systems, also to the operators. The advantage of transparency is that if something goes wrong or if undesired effects arise, it is easier to trace things back to a certain component or mechanism. In that way, the undesired effect does, in many instances, become easier to combat. In a sociotechnical system, transparency also increases the insight that actors have into their contribution to certain effects. In that way, it becomes easier to establish who is responsible for what, and the actors will also feel more responsible because it will be clearer to them just how their actions contribute to certain effects.

Even though a designer can anticipate different unintended and, as yet, undetermined consequences, (s)he cannot possibly also prevent all those negative consequences from arising. From that point of view, technological development always has an experimental character: some social consequences of a given technology only emerge when that technology is implemented. This is partly down to the fact that society also often changes when that technology is embedded; technology and society codevelop as it is phrased.

In creating their own environment, human beings are thus, in essence, experimenting beings. There can be quite a lot at stake in these experiments though. One need only think in this connection of CO_2 emissions, which are contributing to what may turn out to be an irreversible climate change situation, the possible consequences of a nuclear war and the possible consequences of human enhancement by intervening with people's physical and psychological constitution.

From the ethical angle, one may question when such experiments are acceptable and when they cease to be acceptable. There are certain philosophers, like Hans Jonas, who put the case that extreme precautions should be exercised with all the kinds of technology that could conceivably threaten the survival of our species.[42] Another proposition is that such experiments are acceptable if those who possibly stand to suffer from their consequences agree to the experiments going ahead.[43] This notion is similar with the principle of 'informed consent' that is widely used in medical practice. It is the idea that whenever people have to undergo surgery or whenever they participate in a medical experiment, they are first informed as fully as possible of all the possible inherent risks and dangers so that they can decide if they want to undergo the surgery or participate in the experiment. In medical practice, informed consent tends to be implemented on an individual basis; in the case of technological development, that would seem to be more difficult because the relevant 'experiments' often involve large groups all at once. A question which then arises is whether all the members of the group in question have to agree or whether the consent of the majority is enough. Further problems arise in the case of technologies that have possible repercussions for future generations. People who have not yet been born cannot, of course, be asked for their informed consent.

[42]Jonas, H. [1984].
[43]See, for instance, Martin and Schinzinger [2005].

7.4   CONCLUSION

Technological development has an experimental character: unintended consequences cannot be entirely predicted or, for that matter, avoided. This is not just something that is caused by our limited knowledge capacity but also by the fact that the unintended consequences are often the result of the actions of many actors within a sociotechnical system. The implications of this observation are that responsibly developing technology is more complicated than was presumed in Chapter 3. This does not mean to say that one can forget all the suggestions made in Chapter 3; in many cases, they are still relevant, even if success is not guaranteed. In this chapter, we have also demonstrated that engineers can anticipate the occurrence of unintended effects by endeavouring to come up with designs that are robust, flexible and transparent. The experimental nature of technology, finally, gives rise to the ethical question of the conditions under which such experiments are morally acceptable.

7.5   A FEW MORE ISSUES

In this chapter, one of the things we have discussed is why the magnitude of a risk does not determine the acceptability of such a risk. This means, for instance, that if you accept a technology with a certain risk while another technology carries lower risk, you do not automatically have to accept the second technology. Now apply the reasons just mentioned to the comparison between the risks attached to car driving and those attached to nuclear energy. Imagine that the risks of car driving are larger than those posed by nuclear energy and that we find the risks attached to car driving morally acceptable. Why then are the risks of nuclear energy not also necessarily morally acceptable?

A second and more difficult question has to do with the acceptability of social experiments in technology. We have briefly referred to the principle of informed consent as one way of dealing with this particular matter. We also presented two practical objections one could have to applying this principle to technological development as a social experiment. You could also ask yourself if the principle is desirable from a moral point of view. Is it not a much too strict principle? If everyone first has to agree, will technological development still be possible? Will all innovation not then grind to a halt? Is there perhaps another moral point of departure for social experiments that may not have this disadvantage?

Bibliography

Akrich, M. (1992) The description of technical objects. In Bijker, W., and Law, J., Eds. *Shaping Technology/Building Society: Studies in Sociotechnical Change*. Cambridge, MA: MIT, pp. 205–224. 43

Baker, L.R. (2007) *The Metaphysics of Everyday Life: An Essay in Practical Realism*. Cambridge: Cambridge University Press. 6

Bijker, W.E. (1995) *Of Bicycles, Bakelites and Bulbs: Towards a Theory of Sociotechnical Change*. Cambridge, MA: MIT. 89, 91, 92, 93

Bimber, B. (1994) Three faces of technological determinism. In Smith, M., and Marx, L., Eds., *Does Technology Drive History?* Cambridge, MA: MIT, pp. 79–89. 87

Bovens, M. (1998) *The Quest for Responsibility: Accountability and Citizenship in Complex Organisations*. Cambridge: Cambridge University Press. 107, 108

Bundesstelle für Flugunfalluntersuchung. (2004) Untersuchungsbericht AX001–1-2/02, May 2004. Available via `http://www.bfu-web.de/nn_53086/EN/Publications/Investigation_20Report/reports__node.html__nnn=true` Available in English via `http://ocw.mit.edu/NR/rdonlyres/Aeronautics-and-Astronautics/16--358JSpring-2005/C2CB50FC-79B4-413B-9EDD-5B87E9B85CB1/0/ueberlingen.pdf` 76

Collingridge, D. (1992) *The Management of Scale: Big Organizations, Big Decisions, Big Mistakes*. London: Routledge. 113

Ellul, J. (1964) *The Technological Society*. New York: Alfred A. Knopf. 86

Fischer, J.M., and Ravizza, M. (1993) *Perspectives on Moral Responsibility*. Ithaca: Cornell University Press. 108

Grunwald, A. (2001) The application of ethics to engineering and the engineer's moral responsibility: perspectives for a research agenda. *Science and Engineering Ethics*, 7(3), 415–428. DOI: 10.1007/s11948-001-0063-1 50

Hansson, S.O. (2007) Safe design. *Techne*, 10, 43–49. 105

Houkes, W., and Vermaas, P.E. (2010) *Technical Functions: On the Use and Design of Artefacts*. Dordrecht: Springer. 7

Hughes, T.P. (1987) The evolution of large technological systems. In Bijker, W.E., Hughes, T.P., and Pinch, T. J., Eds. *The Social Construction of Technological Systems: New Directions in the Sociology and History of Technology*. Cambridge, MA: MIT, pp. 51–82. 88

Hunter, Th. A. (1997) Designing to codes and standards. In Dieter, G.E, and Lampman, S., Eds. *ASM Handbook*, pp. 66–71. 49

Ihde, D. (1979) *Technology and Praxis: A Philosophy of Technology*. Dordrecht: Reidel. 61

Ihde, D. (1991) *Instrumental Realism: The Interface between Philosophy of Science and Philosophy of Technology*. Bloomington, IN: Indiana University Press. 61

Jonas, H. (1984) *The Imperative of Responsibility: In Search of an Ethics for the Technological Age*. Chicago: University of Chicago Press. 114

Kroes, P. (1996) Technical and contextual constraints in design: an essay on determinants of technological change. In Perrin, J., and Vinck, D., Eds. *The Role of Design in the Shaping of Technology*, Vol. 5, *European Research Collaboration on the Social Shaping of Technology*. pp. 43–76. 95

Kroes, P., and Meijers, A. (2006) The dual nature of technical artefacts. *Studies in History and Philosophy of Science*, 37, 1–4. DOI: 10.1016/j.shpsa.2005.12.001 20

Ladkin, P.B. (2004) Causal analysis of the ACAS/TCAS sociotechnical system. In *ACM International Conference Proceeding Series*, Vol. 110, *Proceedings of the 9th Australian Workshop on Safety Critical Systems and Software*, Vol. 47, pp. 3–12. Available via `http://portal.acm.org/citation.cfm?id=1082339&dl=ACM&coll=GUIDE` 76

Latour, B. (1987) *Science in Action*. Cambridge, MA: Harvard University Press. 61

Latour, B. (1992) Where are the missing masses? In Bijker, W., and Law, J., Eds. *Shaping Technology/Building Society: Studies in Sociotechnical Change*. Cambridge, MA: MIT, pp. 225–258. 43

Martin, M.W., and Schinzinger, R. (2005) *Ethics in Engineering*, 4th ed. Boston: McGraw-Hill. 114

McCormick, B.W. (1979) *Aerodynamics, Aeronautics, and Flight Mechanics*. New York: Wiley. 58

Mitcham, C. (1994) *Thinking through Technology: The Path between Engineering and Philosophy*. Chicago: University of Chicago Press. 1

Pahl, G., and Beitz, W. (1996) *Engineering Design: A Systematic Approach*. London: Springer. 29

Pinch, T.J., and Bijker, W.E. (1987) The social construction of facts and artifacts: or how the sociology of science and the sociology of technology might benefit each other. In Bijker, W.E., Hughes, T.P., and Pinch, T.J., Eds., *The Social Construction of Technological Systems: New Directions in the Sociology and History of Technology*. Cambridge, MA: MIT, pp. 17–50. 89, 91

Pitt, J.C. (2000) *Thinking about Technology: Foundations fo the Philosophy of Technology*. New York: Seven Bridges. 16

Polanyi, M. (1966) *The Tacit Dimension*. London: Routledge & Kegan Paul. 63

Roozenburg, N.F.M., and Eekels, J. (1995) *Product Design: Fundatmentals and Methods*. New Yorks: Wiley. 25

Ryle, G. (1949) *The Concept of Mind*. London: Hutchinson. 63

Schön, D.A. (1983) *The Reflective Practitioner*. New York: Basic Books. 32, 56

Schön, D.A. (1992) Designing as reflective conversation with the materials of a design situation. *Research in Engineering Design*, 3, 131–147. DOI: 10.1007/BF01580516 32

Simon, H.A. (1967) *The Sciences of the Artificial*. Cambridge, MA: MIT. 31, 56

Stone, R.B., and Wood, K.L., (2000) Development of a functional basis for design. *Journal of Mechanical Design*, 122, 359–370. DOI: 10.1115/1.1289637 29

Thomasson, A. (2007) Artifacts and human concepts. In Margolis, E., and Laurence, S., Eds. *Creations of the Mind: Theories of Artifacts and Their Representation*. Oxford: Oxford University Press, pp. 52–71. 6

Van de Poel, I. (1998) *Changing Technologies. A Comparative Study of Eight Processes of Transformation of Technological Regimes*. Ph.D.-Thesis. Mumford series vol. 1.Enschede, Twente University Press. Available via http://www.ethicsandtechnology.eu/images/uploads/Dissertation_Ibo_van_de_Poel.PDF 46

Van der Plas, R. (1999) *Handboek Fietstechniek: Keuze, Reparatie en Onderhoud van de Moderne Fiets*. Rijswijk: Elmar. 90

Van Gorp, A. (2005) *Ethical Issues in Engineering Design: Safety and Sustainability*. Dissertation, Delft University of Technology. Available via http://www.ethicsandtechnology.eu/images/uploads/VanGorp_EthicalIssues_EngineeringDesing_SafetySustainability.pdf 49

Van Inwagen, P. (1990) *Material Beings*. Ithaca: Cornell University Press. 6

Verbeek, P.-P. (2005) *What Things Do: Philosophical Reflections on Technology, Agency and Design*. Penn State: Penn State University Press. 43

Vincenti, W.G. (1990) *What Engineers Know and How They Know It: Analytical Studies from Aeronautical History*. Baltimore: John Hopkins University Press. 27, 57, 60

Weyer, J. (2006) Modes of governance of hybrid systems: the mid-air collision at Ueberlingen and the impact of smart technology. *STI-Studies*, 2, 127–149. Available via http://www.sti-studies.de/fileadmin/articles/weyer-011206.pdf 72, 76

Winner, L. (1992) *Autonomous Technology: Technics-out-of-Control as a Theme in Political Thought.* Cambridge, MA: MIT. 85, 86

Authors' Biographies

MAARTEN FRANSSEN

Maarten Franssen has degrees in theoretical physics and history and obtained a doctorate in philosophy on issues in the foundations of the social sciences. He joined the Department of Philosophy at Delft University of Technology in 1996. His main areas of research are conceptions of normativity and rationality in general, the application of rational decision-making methods in scientific and technical research, and the formal description of artefacts and systems in technology.

WYBO HOUKES

Wybo Houkes is a senior lecturer in the Philosophy of Science and Technology Department at Eindhoven University of Technology. He studied theoretical physics in Amsterdam and philosophy in Leiden, where gained his doctorate on Kantianism in early 20th century philosophy. He is currently conducting research into the tension between universal Darwinism and more tailor-made approaches in evolutionary theories of technology, the role of agency in these theories, and into the nature of technological knowledge and the functions of technical artefacts.

PETER KROES

Peter Kroes studied technological physics at Eindhoven University of Technology and did a doctorate on the philosophical problems surrounding the concept of time in the field of physics at Radboud University Nijmegen. He holds the chair of philosophy and ethics at Delft University of Technology where he lectures on the philosophy of science and technology. His main fields of research are the nature of technical artefacts and their role in sociotechnical systems, and the philosophy of technical designing.

IBO VAN DE POEL

Ibo van de Poel is associate professor in ethics and technology at Delft University of Technology. He studied philosophy of science, technology and society at Twente University, where he did a PhD in science and technology studies. He teaches engineering ethics and does research into design and values, the moral acceptability of technological risks, responsibility in research and development networks, the ethics of new emerging technologies, and the notion of technology as a form of social experimentation.

PIETER VERMAAS

Pieter Vermaas studied theoretical physics at the University of Amsterdam and gained a doctorate from Utrecht University on the philosophy of quantum mechanics. Since 1998, he has been affiliated to the Department of Philosophy at Delft University of Technology where he is doing research into the principles of technology. His subjects of interest within that field are the analysis of the concept of technical function as used within engineering, and the description of designing as given by the various design methodologies.

Index

Printed in the United States
by Baker & Taylor Publisher Services